Educational Neuroscience

Edited by
Kathryn E. Patten and Stephen R. Campbell

WILEY-BLACKWELL

A John Wiley & Sons, Ltd., Publication

This edition first published 2011
Originally published as Volume 43, Issue 1 of *Educational Philosophy and Theory*
Chapters © 2011 The Authors
Book compilation © 2011 Philosophy of Education Society of Australasia

Blackwell Publishing was acquired by John Wiley & Sons in February 2007. Blackwell's publishing program has been merged with Wiley's global Scientific, Technical, and Medical business to form Wiley-Blackwell.

Registered Office
John Wiley & Sons Ltd, The Atrium, Southern Gate, Chichester, West Sussex, PO19 8SQ, United Kingdom

Editorial Offices
350 Main Street, Malden, MA 02148-5020, USA
9600 Garsington Road, Oxford, OX4 2DQ, UK
The Atrium, Southern Gate, Chichester, West Sussex, PO19 8SQ, UK

For details of our global editorial offices, for customer services, and for information about how to apply for permission to reuse the copyright material in this book please see our website at www.wiley.com/wiley-blackwell.

The right of Kathryn E. Patten and Stephen R. Campbell to be identified as the author of the editorial material in this work has been asserted in accordance with the UK Copyright, Designs and Patents Act 1988.

Wiley also publishes its books in a variety of electronic formats. Some content that appears in print may not be available in electronic books.

Designations used by companies to distinguish their products are often claimed as trademarks. All brand names and product names used in this book are trade names, service marks, trademarks or registered trademarks of their respective owners. The publisher is not associated with any product or vendor mentioned in this book. This publication is designed to provide accurate and authoritative information in regard to the subject matter covered. It is sold on the understanding that the publisher is not engaged in rendering professional services. If professional advice or other expert assistance is required, the services of a competent professional should be sought.

Library of Congress Cataloging-in-Publication Data

Educational neuroscience / edited by Kathryn E. Patten, Stephen R. Campbell.
    p. cm. – (Educational philosophy and theory special issues)
    Includes bibliographical references and index.
    ISBN 978-1-4443-3985-7 (pbk.)
    1. Educational psychology.   I. Patten, Kathryn E.   II. Campbell, Stephen R.
    LB1501.E38 2011
    370.15–dc22

                                                              2011013766

A catalogue record for this book is available from the British Library.

This book is published in the following electronic formats: ePDFs 9781444345797; Wiley Online Library 9781444345827; ePub 9781444345803; Kindle 9781444345810

Set in 10pt Plantin by Toppan Best-set Premedia Limited
Printed in Malaysia by Ho Printing (M) Sdn Bhd

01   2011

Educational Neuroscience

Educational Philosophy and Theory Special Issue Book Series

Series Editor: Michael A. Peters

The *Educational Philosophy and Theory* journal publishes articles concerned with all aspects of educational philosophy. Their themed special issues are also available to buy in book format and cover subjects ranging from curriculum theory, educational administration, the politics of education, educational history, educational policy, and higher education.

Titles in the series include:

Contents

Notes on Contributors

Daniel Ansari is an Associate Professor of Developmental Psychology and the Canada Research Chair in Developmental Cognitive Neuroscience at the University of Western Ontario. His primary interest is in the neurocognitive trajectories underlying the development of typical and atypical numerical and mathematical skills. He uses both behavioural and brain imaging methods to better understand how children develop numerical skills and what neuronal mechanisms underlie the development of mathematical competencies. Email: daniel.ansari@uwo.ca

Stephen R. Campbell is Associate Professor and Director of the Educational Neuroscience Laboratory <www.engrammetron.net> in the Faculty of Education at Simon Fraser University. His scholarly focus is on the historical and psychological development of mathematical thinking from an embodied perspective informed by Kant, Husserl, and Merleau-Ponty. His research incorporates methods of psychophysics and cognitive neuroscience as a means for operationalizing affective and cognitive models of math anxiety and concept formation. Email: sencael@sfu.ca

Donna Coch is an Associate Professor in the Department of Education at Dartmouth College. Using a combination of behavioural measures and a noninvasive brain wave recording technique, her research focuses on the reading brain. A goal of both her research and her teaching is to make meaningful connections across mind, brain, and education. Email: donna.coch@dartmouth.edu

Michel Ferrari teaches developmental and educational psychology at the Ontario Institute for Studies in Education at the University of Toronto. His most recent co-edited book is *Developmental Relations Among Mind, Brain, and Education: Essays in Honor of Robbie Case* (Springer, 2010, with Ljiljana Vuletic). In 2010, he also edited a special issue of the *History of the Human Sciences* on the history of the science of consciousness and is preparing a *Handbook on Resilience in Children of War* (Springer, in press, with Chandi Fernando). He is currently leading an international study on the personal experience of wisdom as part of a general program of research into the importance of personal development for quality of life. Email: michel.ferrari@utoronto.ca

Kurt Fischer leads an international movement to connect biology and cognitive science to education, and is founding editor of the journal *Mind, Brain, and Education* (Blackwell), which received the award for Best New Journal by the Association of American Publishers. As Director of the Mind, Brain, and Education Program and Charles Bigelow Professor at the Harvard Graduate School of Education, he does research on cognition, emotion, and learning and their relation to biological development and educational assessment. In his research he has discovered a general scale that provides tools for assessing learning and development in any domain. His most recent books include *The Educated Brain* and *Mind, Brain, and Education in*

Reading Disorders (Cambridge University Press, 2008 and 2007, respectively). Email: kurt_fischer@harvard.edu

John Geake is Professor of Learning and Teaching and Deputy Head of School, School of Education, University of New England, Australia, where his research has concentrated on applications of neuroscience to children's learning. Prior to taking up this position in 2009, Professor Geake was Professor of Educational Neuroscience, Oxford Brookes University, Oxford UK, where his work focussed on applications of neuroscience to educational outcomes. Email: jgeake@une.edu.au

Jeanne Marcum Gerlach is Associate Vice President for K-16 Initiatives and Dean of the College of Education and Health Professions at the University of Texas Arlington. Her research focuses on Urban Education, Business/Higher Education Partnerships, Issues in English Education, Writing As Learning, Women in Leadership Roles, Collaborative Learning, and Governance in Higher Education. She is the coeditor of *Missing Chapters: Ten Pioneering Women In NCTE and English Education* and co-author of the book, *Questions of English: Ethics, aesthetics, rhetoric, and the formation of the subject in England, Australia and the United States.* Dr Gerlach has taught in England, New Zealand, France, Germany, Thailand, and Australia. Her awards include the National Council Teachers of English Outstanding Woman In English Education and the University of North Texas' and West Virginia University's Outstanding Alumni Award. She received the Fort Worth Business Press Great Women of Texas Most Influential Woman Award, 2002. Email: gerlach@uta.edu

Paul Howard-Jones is Senior Lecturer at the Graduate School of Education, University of Bristol. His research focuses exclusively on issues interfacing neuroscience and education. He publishes in neuroscience, psychology and education and coordinates the Neuroeducational Network (NEnet: www.neuroeducational.net). His latest book is *Introducing Neuroeducational Research* (Routledge, 2010). Email: paul.howard-jones@bris.ac.uk

Mary Helen Immordino-Yang, EdD is a social/affective neuroscientist and educational psychologist who studies the brain bases of emotion, social interaction and culture and their implications for development and schools. She is an Assistant Professor of Education at the Rossier School of Education and an Assistant Professor of Psychology at the Brain and Creativity Institute, University of Southern California, and Associate Editor for North America of the journal *Mind, Brain and Education.* A former junior high school teacher, she earned her doctorate in human development at Harvard University, and completed her postdoctoral training in affective neuroscience with Antonio Damasio. She was the inaugural recipient (2008) of the Award for Transforming Education through Neuroscience, cosponsored by IMBES and the Learning and the Brain Conference, and lead author of a 2009 Cozzarelli Award-winning paper, sponsored by the Editorial Board of the Proceedings of the National Academy of Sciences. Email: immordin@usc.edu

Anthony E. Kelly is Professor of Educational Psychology at George Mason University. He has published a number of articles related to educational research methods, and is editing a volume on the neural basis for mathematics learning. Dr Kelly has a number of grants from the US National Science Foundation, and is a New Century Scholar in the Fulbright Program. Email: akelly1@gmu.edu

Hideaki Koizumi is a Fellow at the Advanced Research Laboratory, Hitachi Ltd. Hatoyama, Japan, and Director of the Research and Development Division of Brain-Science & Society at the Research Institute of Science and Technology for Society, Japan Science and Technology Agency. He is a Visiting Professor, Research Center for Advanced Science and Technology, The University of Tokyo. He has been advocating the concept of trans-disciplinarity since 1995, and been leading a new field of applied brain science including brain-science and education. He has also developed various noninvasive brain imaging technologies, such as MRI, fMRI and fNIRS (Optical Topography). Email: hideaki.koizumi.kd@hitachi.com

Kerry Lee is an Associate Professor of Psychology at the National Institute of Education, Singapore. He has interests in the application of laboratory-based findings to various forensic and educational issues. In recent years, he has focused on individual differences in mathematical proficiency. Using both experimental and correlational methods, he and his colleagues have examined the contributions of working memory and executive functioning to children's performances on algebraic word problems. He is also interested in the use of neuroimaging techniques to examine pedagogically relevant questions. Email: Kerry.Lee@nie.edu.sg

Fenna van Nes recently completed her PhD at the Freudenthal Institute for Science and Mathematics Education in Utrecht, the Netherlands. She has published several articles about young children's spatial structuring ability and the development of early spatial sense and number sense. In her thesis she describes the design of a series of lesson activities that she developed, which can be performed in kindergarten classrooms to foster children's mathematical development. Email: fennavannes@gmail.com

Swee Fong Ng is Associate Professor with the Mathematics and Mathematics Education academic group at the National Institute of Education, Nanyang Technological University, Singapore. Prior to joining the National Institute of Education, she spent about twenty years in Malaysia teaching mathematics at the upper secondary level. She now works extensively with both pre-service primary mathematics teachers as well as in-service mathematics teachers. Her other responsibilities include teaching and supervising at the master and doctoral level. Her general interest is looking at ways to help improve the teaching and learning of mathematics across the curriculum. The teaching and learning of algebra is her special interest. Email: sweefong.ng@nie.edu.sg

Kate Patten is the Outreach Coordinator for ENGRAMMETRON, the Educational Neuroscience Laboratory at Simon Fraser University. Kate's current research interests lie in the neuroscience and neuropsychology of emotion and its implications for neuropedagogy, specifically within the research field of educational neuroscience. She is also interested in the role of emotion regulation in the classroom, as well as the debunking of myths encountered in 'brain-based education'. Email: kepatten@sfu.ca

Marc Schwartz is Professor of Mind, Brain and Education at the University of Texas, Arlington (UTA), and president-elect of the International Mind, Brain and Education Society (IMBES). He is also director of the recently established Southwest Center for Mind, Brain and Education at UTA, The center seeks to identify and support promising research agendas at the intersection of neuroscience and cognitive science to inform educational practice and leadership. His research focuses on how the dynamic enterprise of learning unfolds, through perspectives ranging from the student's to the institutions that oversee the student's learning. Email: schwarma@uta.edu

Bert De Smedt is an Assistant Professor of Educational Neuroscience at the Katholieke Universiteit Leuven, Belgium. He has published a number of articles related to the neurocognitive correlates of individual differences in mathematical achievement. He has done work on mathematical performance in developmental disorders, including dyscalculia, dyslexia, and genetic disorders. He is particularly interested in making connections between education and neuroscience. Email: Bert.DeSmedt@ped.kuleuven.be

Zachary Stein EdM is currently a doctoral candidate at Harvard in the Mind, Brain, and Education department. He has published on topics in the philosophy of education, neuroscience, interdisciplinarity, developmental psychology, and psychometrics, in journals such as *American Psychologist*, *New Ideas in Psychology*, and *Journal of Philosophy of Education*. Zak is also the Deputy Director of Development Testing Service, Inc. (DTS), a non-profit research and development organization that focuses on building usable knowledge and technology at the interface of psychometrics, test design, developmental psychology, and education. E-mail: stein.zak@gmail.com

Foreword

The *Educational Philosophy and Theory* Book Series is dedicated to enhancing the ongoing conversations surrounding all aspects of educational philosophy, including areas of pure and applied educational research. The book series aims to extend the dialogues of educational philosophy by incorporating work from the related fields of arts and sciences, as well as work from professional educators. This monograph based on the special issue entitled *Educational Neuroscience* and edited by Kathryn Patten and Stephen Campbell brings together fourteen chapters, including an Introduction, to review and discuss an emerging field sometimes also referred to as Mind Brain Education (MBE), after the journal established by Kurt Fischer in 2007. Both Kate Patten and Sen Campbell are from the Educational Neuroscience Laboratory (respectively, Outreach Coordinator and Director) established at Simon Fraser University in 2006 through the Canadian Foundation for Innovation's New Opportunities Program. The Laboratory called Engrammetron, after the 'engram' or 'memory traces' hypothesized by Karl Lashley (1890–1958) the father of modern neuroscience, was set up with a primary specialization in mathematics education as a facility to measure, analyze and observe through various instruments and methods (including, electroencephalography (EEG), electrocardiography (EKG), electromyography (EMG), and eye-tracking (ET) capability), patterns of 'mind brain' behaviour. The field is very recent and emerging quickly with major centres or research networks established in London, Cambridge, Harvard and Bristol:

- London (Centre for Educational Neuroscience, http://www.educationalneuroscience. org.uk/)
- Cambridge (Centre for Neuroscience in Education, http://www.educ.cam.ac.uk/ centres/neuroscience/)
- Harvard (Brain Mind, and Education, http://www.gse.harvard.edu/academics/masters/ mbe/)
- Bristol (The NeuroEducational Research Network, http://www.neuroeducational. net/)

All established in the past five years, these facilities advertise themselves as transdisciplinary projects designed to synthesize biological, cognitive and social dimensions of learning within a developmental psychology framework that pays homage to Piaget. The Cambridge Centre states 'we aim to understand how the brain functions and changes during the development of reading and maths, exploring the development of related skills such as language, memory, numerosity and attention'. The Harvard initiative advertises an interdisciplinary programme 'including not only psychology, pedagogy, and neuroscience, but also philosophy, anthropology, linguistics, computer science, and other relevant disciplines.' The Centre for Educational Neuroscience at London, an inter-institutional project of University College London, the Institute of Education and Birkbeck College, on its website records conference presentations for 'Educational Neuroscience: An Emerging Discipline' held at Birkbeck in June 2010 with papers on Individual differences in numerical and mathematical abilities, the social brain in adolescence, aspects of numeracy and math learning disability, school science, language and literacy, as well as autism and dyslexia.

In addition, there also exist various SIGS and forums. Most organizations and educational neuroscientists tend to picture themselves as providing a link between biology and cognition; many also acknowledge links to other disciplines, including philosophy and technology. In his scoping chapter Sen Campbell pictures educational neuroscience as a new area of educational research that goes beyond a conception of applied cognitive neuroscience. Drawing on a theory of the embodied mind put forward in the early 1990s by Francisco Varela and his colleagues who sought to overcome the Cartesian Anxiety by complementing cognitivism as an outgrowth of cybernetics with emergence or connectionism, Campbell focuses on subjective experience to argue 'any changes in subjective experience must in principle manifest objectively in some manner as changes in brain, body, and behaviour, and vice versa' (pp. 9–10).

What I like about Campbell's conception is that it is based on philosophical commitments and a good working knowledge of philosophy of mind which makes it both suitable and highly relevant for our readers and for its inclusion in the *Educational Philosophy and Theory* book series.

I am grateful to Kate Patten and Stephen Campbell for their editorial work in bringing such an excellent international collection together from leading scholars in this rapidly emerging field, themselves included. Educational neuroscience promises new characterizations of the learner in terms of brain, genetic and hormonal states; its applications in mathematics, literacy and social or emotional cognition are interesting even although it still faces formidable methodological and philosophical challenges; and yet already it has already accomplished important work such as deconstruction of prevalent neuromyths such as left/right or male/female brain.

Michael A. Peters
University of Illinois

1
Introduction: Educational Neuroscience

Kathryn E. Patten & Stephen R. Campbell

This book provides an overview of a number of recent initiatives in a new area of research that is coming to be known as educational neuroscience. Educational neuroscience, as a first approximation, variously involves syntheses of theories, methods, and techniques of the neurosciences, as applied to and informed by educational research and practice. Contributions to this book were sought from principals involved in initiatives pertaining to educational neuroscience with common foci on 1) motivations, aims and prospects; 2) theories, methods, collaborations; and 3) challenges, results, and implications, both potential and actual, resulting from these initiatives. Contributors were asked to write position statements with special emphasis on the motivations, methodologies, and practical implications of their particular initiatives for educational philosophy and theory, as well as for educational research and pedagogy.

What emerges in this book is an indication of the wide range of initiatives related to educational neuroscience. This book presents a wide variety of initiatives and methodologies, as well as common goals, concerns and issues. Many topics raised herein are endemic to the emergence of a new discipline: for instance, a need for more coherent terminology, a struggle to identify and establish theoretical and philosophical foundations, a quest for practical empirically-based models, and a requirement for standards of ethical practice. Amplifying problems in establishing the new discipline of educational neuroscience is its cross-disciplinary nature and its consequential need to combine a variety of resources, methodologies, and results. In order to include as wide a variety of responses as possible, authors truncated their submissions to present brief overviews of their perspectives, purposes, portents, and projects. The authors examine a variety of concerns, issues, and directions relating to educational neuroscience; as well as revealing a need to establish theories, models, ethics, methodologies and a common language.

Stephen Campbell, an educational philosopher and researcher in mathematics education at Simon Fraser University, opens this book by considering the nature of educational neuroscience. In so doing, he identifies its proper object of study as the 'mindbrain'. Campbell advocates a radical theory of embodied cognition that takes as a foundational assumption that any and all changes in subjective experience necessarily entail associated changes in brain and body behavior. Accordingly, he has been expanding his empirical research in mathematics education to include methods and techniques of psychophysiology and cognitive neuroscience in his studies of mathematical cognition and learning.

Educational Neuroscience, First Edition. Edited by Kathryn E. Patten and Stephen R. Campbell.

In Chapter 3, Anthony (Eamonn) Kelly, Professor and Coordinator of Instructional Technology at George Mason University, identifies many relevant factors contributing to educators' growing interest in the findings of cognitive neuroscience. He asserts that neuroscience may well provide the empirical 'primitives' for theorizing anew about learning; in fact, spawning a revolution in our understanding of learning grounded in science. He emphasizes the need to debunk brain-based neuromythologies and replace vague theories of learning with mixed method research-based theories involving a range of disciplines, including the neurosciences. These new theories that incorporate empirical research will ground changes in pedagogy, as in such collaborations as Science, Technology, Engineering, and Mathematics (STEM) learning. STEM seeks to define fundamental aspects of learning based on neural processes and other biological foundations, and in so doing, to aid in clarifying, defining, and creating theories and models of learning. As well, he argues, STEM has a role to play in helping to establish research agendas and in disseminating resultant findings to various disciplines contributing to educational neuroscience.

In Chapter 4, Paul Howard-Jones, Senior Lecturer at the University of Bristol, argues that it is imperative to include brain function in current educational theorizing. He cautions that collaborations between neuroscience and education are fraught with philosophical, conceptual, methodological, and practical issues. He also cautions against 'medicalising' educational issues in our quest for understanding educational issues, and presents a 'levels-of-action-model' that incorporates the brain-mind-behaviour paradigm as a workable interface of the natural and social sciences by neuroeducational researchers. Specifically, Howard-Jones presents and discusses the Neuroeducational research network (NEnet) at the University of Bristol in its role to develop collaboration between the fields of neuroscience and education.

What follows in Chapter 4 is Michel Ferrari's view of educational neuroscience as 'an exciting renovation' of cognitive neuroscience and other neurosciences that will advance our understanding of how knowledge and cognition is embodied. Michel, Head of the Centre for Applied Cognitive Science at the Ontario Institute of Studies in Education, advocates that while neuroscientific investigation typically addresses pathologies of learning disabilities, our focus as educational researchers should be to understand the larger underlying context of personal learning and development and to avoid neuroscientific labeling of atypical students in manners that are limiting and potentially stigmatizing. He cautions against an all-too-common practice of over-generalizing laboratory results to learning situations in situ, and against acceptance of frameworks that negate the presence and importance of agency. Ferrari argues, in some contrast to Howard-Jones, that educational strategy must follow the medical model in that pure research informs practice. Concomitantly, he argues that this strategy must also be socially imbedded and culturally mindful in that it reflects the values we espouse and the society to which we aspire.

Daniel Ansari, and associates Donna Coch (Dartmouth College) and Bert de Smedt (Katholieke Universiteit Leuven) of his Numerical Cognition Laboratory at the University of Western Ontario, examine the role of cognitive neuroscience in informing education. They acknowledge that changing educational theories and models to be neuroscientific and grounded in biology will be complex and necessarily involves changes in teacher education and teacher training. Advocating that cognitive neuroscientists take an essential role in helping teachers to become literate in neuroscience, they concomitantly propose that teachers reciprocate by enabling cognitive neuroscientists to become literate in the issues and problems related to classroom practice. This process would replace the application of the myths of brain-based learning with interdisciplinary applied research and would generate new collaborations, new paradigms, and eventually, changes in pedagogy.

In Chapter 7, John Geake of the School of Education, University of New England, regards educational neuroscience as an interdisciplinary field both inspiring and inspired by educators' questions pertaining to pedagogy and curriculum arising from educational problems and issues. To this end, he espouses the use of neuroscientific action research to both validate some current pedagogical practices and to provoke some new ones. Geake laments the lack of recognition of the function of the human brain in most education policy, curriculum and outcome documents. For Geake, it is the job of educational neuroscience to include brain function in education. Educational neuroscience, he asserts, needs its own discipline-specific methodology that addresses the issues, concerns, problems, and needs of educators and learners, but at the same time embraces the findings and expertise of cognitive neuroscientists. Geake briefly presents his research on fluid analogy-making as a basic cognitive process underlying creative thinking. These brain functions, such as analogy making, can be empirically validated using such instruments as fMRI. Geake and his colleagues have proposed a neuropsychological model of creative intelligence that features fluid analogizing.

As well as ascertaining the need for educational neuroscience, undertaking the task of defining it, and establishing its place in the realm of educational research, other contributors to this book address the problem of how research is to be conducted.

Hideako Koizumi, Director of the Research Institute of Science and Technology for Society in Japan, welcomes the biologically-grounded perspective of educational neuroscience. He regards learning as making neuronal connections in response to external stimuli from the environment, and education as the process of creating and/or controlling stimuli, as well as 'inspiring the will to learn'. He discusses the use of longitudinal cohort studies using twins that chart the development of brain function with regard to environmental and genetic factors. He argues that such studies will enable researchers to contribute to educational policy making, reveal potential effects of technology, and help determine whether findings from animal studies can be applied to humans. Neatly summarizing several cohort studies, their objectives, and their methodologies, Koizumi presents the advantages of cohort studies, as well as possible issues and implications.

While there is agreement that multidisciplinary collaboration is needed, Zachary Stein and Kurt Fischer at Harvard University Graduate School in Education, next propose a model for the training of a new generation of educational researchers and practitioners in neuroscience. They present the idea of research school collaborations as the model of choice for Mind, Brain, and Education (MBE). They argue that research school collaborations embody the methodological innovations necessary to build a functional interdisciplinary research group. As well, Stein and Fischer identify important issues for MBE: the control of quality and interdisciplinary synthesis of methods; the development of pragmatic, comprehensive models of human development; the need to develop ethics that govern the use of neuroscientific research findings; and the need to create a common lexicon. They advocate problem-based research in the complex context of practice, involving methodological pluralism, both quantitative and qualitative analyses, with the goal of improving pedagogy.

Marc Schwartz and Jeanne Gerlach, at the Southwest Center for Mind, Brain and Education at the University of Texas at Arlington, further along the lines of Stein and Fischer, describe the reincarnation of Dewey's laboratory school, a network of researchers, educators, administrators, and policy makers working collaboratively in what they term a Research Schools Network. This network is established to provide the forum for establishing conceptual frameworks, identifying educational challenges, developing experimental methodologies and ethics, clarifying research findings, interpreting conclusions, and monitoring suitable applications of results. Rather than call the new field educational neuroscience, they prefer the term Mind, Brain and Education, which they see as being more pedagogically focused. MBE shares the vision of educational neuroscientists: to improve our understanding of learning and to actualize this knowledge in pedagogy that reflects the multidisciplinary perspective of the mind, including planning, teaching, and assessment.

While many have been theorizing about the new field of educational neuroscience, asserting its place both in neuroscience and education, and examining the creation of research communities and their practice, others have forged ahead and used established methodology to apply to the examination of specific learning tasks. Some of these initiatives are particularly evident in the area of mathematics education research.

Fenna van Nes, in Chapter 11, discusses the Mathematics Education and Neurosciences (MENS) Project at the Freudenthal Institute for Science and Mathematics Education. Van Nes advocates bidirectional collaboration between mathematics education researchers and neuroscientific researchers, with a view to improve children's mathematical learning. She sees neuroscientific research and educational research not as a fusion of fields, but as an interdisciplinary sharing of insights. She and her colleagues combine qualitative 'design research' with quantitative 'experimental research' to arrive at a more comprehensive understanding of the prerequisites involved in the development of early spatial structuring and patterning ability in order to relate this early learning to later mathematical performance. While the mathematics education researchers examine the role of kindergarteners' spatial structuring ability, the neuroscience researchers study the kindergarteners' automatic quantity processing and its consequence on mathematical development. This combined knowledge of the sharing of these findings, she projects, will lead to better educational practice in the arenas of diagnosis, prevention, and intervention in the learning and teaching of mathematics.

Kerry Lee and Swee Fong Ng at the National Institute of Education in Singapore, also focus on neuroscience and the teaching of mathematics. They aim to differentiate among the neuroanatomical brain systems utilized for doing math, teaching math, and learning math. Advocating a mixed method approach to problem solving, they also raise the issue of transferring laboratory research findings to the real classroom and the imperative of pragmatic research in order to extend legitimate laboratory findings to pedagogy. Challenges they address include a condensation of issues related to teaching and learning mathematics into tasks suitable to the constraints of neuroimaging techniques, as well as the issues of drawing legitimate inferences for pedagogy from the research results. Specifically, Lee and Ng utilize neuroimaging techniques to investigate heuristics, namely the model method versus formal algebra, utilized in teaching algebraic word problem solving as well as to investigate the sequential steps in problem solving of progressing from model or equation to solution. Their neuroimaging studies have proven useful in providing insights into developing suitable interventions for improving students' problem solving success in algebra. On the practical side, professional development courses are offered to share with teachers how they can improve their pedagogy to enhance mathematical learning.

Having predicated her doctorate with many years of classroom experience, Kathryn Patten, Outreach Coordinator of the Educational Neuroscience Laboratory at Simon Fraser University, believes that the education curriculum has detrimentally ignored the emotional needs of children. Arguing for the primal phylogenetic function of emotion over cognition, she examines both the neuroscience and neuropsychology of emotion to present the Somatic Appraisal Model of Affect (SAMA). Based upon Damasio's somatic marker theory and borrowing from Lazarus' appraisal theory, Patten differentiates among three levels of affect: dispositions or moods; basic universal, instinctive emotions; and feelings, both secondary and conscious. Feelings, she posits, involve cognitive appraisal regarding goals, cultural practices, beliefs, and a sense of self. SAMA is presented as a dynamic model to provoke change in how educators and researchers regard emotion and upon which to investigate possible changes in policy, curricula, and practice that will address the emotional needs of students.

Mary Helen Immordino-Yang, at the Brain and Creativity Institute and Rossier School of Education at the University of Southern California, is committed to bringing neuroscientific evidence to inform educational theory and practice. As an educator, she regards our collective role in reconciling new neuroscientific findings with established theories, and uncovering how this new knowledge may be used to improve teaching and learning. Immordino-Yang places the embodied mind in the context of the *polis*, arguing educational neuroscientists must reconcile theories on which good practice is based with new neuroscientific evidence of mind/body functions that incorporate the foundations of development and involving emotional and social learning. Theories must incorporate research on emotion and social processing, as these functions modulate neural processing and, hence, learning. Immordino-Yang welcomes these imminent changes in educational theory, models, and practice evoked by educational neuroscience and anticipates that findings in affective and social neuroscience will have a profound impact on our understanding of development and learning. The evolution of educational theory and the resultant models will lay the groundwork for changes in educational practice and have implications for the design of new learning environments.

In conclusion, we wish to thank the authors of this book for their contributions. These chapters represent a limited number of snap-shots of a very rapidly developing field. There are many other initiatives than have been indicated here. We hope this special edition of *Educational Philosophy and Theory* on Educational Neuroscience will evoke discussion, prompt exploration, inspire research, and add spark to the continued emergence of this new, exciting field that holds promise to transform education as we know it. Implicit in collaborations that constitute and complement educational neuroscience is the challenge of accommodating various theoretical and philosophical stances of diverse disciplines. Despite these differences, initiatives in educational neuroscience share a common aim: to produce results that ultimately improve teaching and learning, in theory and in practice. It is our hope that this book will provide one of an increasing number of forums that will help document and facilitate the voyage of philosophers, theorists, researchers, and practitioners into this exciting new millennium of educational neuroscience.

2
Educational Neuroscience: Motivations, methodology, and implications

STEPHEN R. CAMPBELL

Introduction

'What does the brain have to do with learning?' *Prima facie*, this may seem like a strange thing for anyone to say, especially educational scholars, researchers, practitioners, and policy makers. There are, however, valid objections to injecting various and sundry neuroscientific considerations piecemeal into the vast field of education. These objections exist in a variety of dimensions. After providing a working definition for educational neuroscience, identifying the 'mindbrain' as the proper object of study thereof, I discuss, dispel or dismiss some of these objections prior to presenting my motivations, aims, and prospects for this new area of educational research. I then briefly outline a positive case for educational neuroscience in terms of theories, methods, and collaborations, and conclude with a brief discussion of some challenges, results, and implications thereof. Naturally, the following considerations are but my own, some of which may be shared to some extent by others working in this area, as the case may be.

Defining Educational Neuroscience

In defining educational neuroscience, I do not presume to be putting forth anything more than an evolving working definition pertaining first and foremost to the approach I have been taking to some of the work I have engaged upon (e.g. Campbell, 2002; 2003; 2004; 2005; 2006a,b; 2007; 2010). My working definition attempts to go beyond thinking of educational neuroscience in a more narrow but quite legitimate sense, as an *applied* cognitive neuroscience. However, educational neuroscience can certainly be perceived and pursued as such, especially if there are no substantive differences in philosophical and methodological orientations to be found between educational and cognitive neuroscience. Moreover, educational neuroscience can also be considered more broadly than the working definition I am putting forth here, as concerning pretty well anything that involves some kind of rigorous (viz., methodological/scientific) synthesis concerning matters pertaining to mind, brain, and education. The term 'neuro-education' encapsulates this latter view quite well (for a comprehensive treatment of these matters, see Tokuhama-Espinosa, 2008). I leave it for others to judge the extent to which my working conception of educational neuroscience relates more closely to a more narrowly defined applied cognitive neuroscience (and, hence, more akin to cognitive neuroscience) or more closely to a broadly defined neuroeducation (that itself, perhaps, is viewed as more akin to an education science).

Educational Neuroscience, First Edition. Edited by Kathryn E. Patten and Stephen R. Campbell.
Chapters © 2011 The Authors. Book compilation © 2011 Educational Philosophy and Theory/Blackwell Publishing Ltd.
Published 2011 by Blackwell Publishing Ltd.

For myself, I see educational neuroscience as a new area of *educational* research, and one that naturally draws on the neurosciences (especially cognitive neuroscience, including psychophysiology), and yet one that falls within the broader emerging *field* of neuroeducation. That is to say, I see educational neuroscience as an area of educational research that draws on, as in being informed by, theories, methods, and results from the neurosciences, but unlike an applied cognitive neuroscience, is *not restricted* to them. More positively, the focal points of educational neuroscience are living human beings, not *just* physiological and biological mechanisms underlying them. As such, in my opinion, educational neuroscience, by definition, with respect to teachers' teaching and learners' learning, must attempt to bridge, or at least come to terms with, the gap between conscious minds and living brains. This, of course, is a hard problem, or, rather, for some, it is *the* hard problem (Chalmers, 1995). There is a well-known explanatory gap between the ontological schism Descartes, and others before him and since, have wrought between mind and the body (viz., brain). Perennial attempts to reduce one to the other (as materialists and idealists have been wont to do) are fraught with difficulty (e.g. Campbell & Dawson, 1995; Velmans, 1995). It is not necessary for educational neuroscientists of various ilks and inclinations to hold identical philosophical views regarding these matters, but I think it does behoove researchers in this area to be open and clear about where they stand on them. Educational neuroscientists, be they cognitive neuroscientists applying their trade to educational problems or educational researchers applying methods and techniques of cognitive neuroscience, can adhere to the same research paradigms, e.g. those associated with brain imaging, despite fundamental philosophical differences that may exist between us (Campbell, forthcoming).

Clearly a synthesis of sorts between education and neuroscience, educational neuroscience can also be viewed variously as a multidisciplinary, interdisciplinary, or transdisciplinary endeavor (cf. Gibbons *et al.*, 1994). My preference is the latter. Educational neuroscience as a multidisciplinary activity would typically involve neuroscientists and educationists (i.e. educational theorists, researchers, practitioners, and policy-makers) contributing their respective expertise to a common project, with little appetite for engaging each other's theories, methods, practices, or policies let alone, philosophical commitments—each expert would essentially just 'do their own thing'. Educational neuroscience as an interdisciplinary activity, however, would find neuroscientists and educationists actively engaging each other's points of view in an attempt to jointly optimize each other's respective contributions to a given project. Educational neuroscience as a bona fide transdisciplinary activity, by definition, must entail the forging of new philosophical frameworks and research methodologies for variously bridging education and neuroscience, mind and brain, phenomenological and physiological, teleological and causal, first person and third person, objective and subjective, and so forth.

Whereas multidisciplinary and interdisciplinary activities are typically project-oriented, based on treating traditional problems in new ways, transdisciplinary activity is more oriented toward opening new, potentially revolutionary, sets of problems. The 'holy grail', for a transdisciplinary educational neuroscience as I see it, would be to empower learners through the volitional application of minds to consciously perceive and alter their own brain processes into states more conducive to various aspects of learning. I anticipate that such a discipline would be transdisciplinary to the extent that it would have to deal with the hard problem of the nature of consciousness and the explanatory gap between mind and matter. So conceived, educational neuroscience would have to engage and come to terms with how mental processes can exercise causal effects on brain processes, and not just be satisfied with understanding how brain processes exercise causal effects on mental processes, the latter as per the wont of neuroscience. This bridging of the subjective with the objective itself may require grounding in quantum physics in an attempt to reconcile a foundational divide between Aristotelian and Newtonian physics, the former according to which objects themselves influence their own motion, and the latter to which the motion of objects change only through application of external forces.

The notion that human beings have it within our potential to consciously alter our brains into states more conducive to various aspects of learning is a tantalizing thought. The significance of cultivating such ability could potentially rival the advent of literacy. No doubt, some might view this 'holy grail' for educational neuroscience not simply as a (few) bridge(s) too far, but as an impossible pipe dream, or worse. Who knows what this 21st century might bring? (Let s/he who is without speculation cast the first critique!) Others may wisely counsel a more cautious approach with less lofty aspirations, and, frankly, I also find myself somewhat of this disposition, at least in the short term. In his landmark article, Bruer (1997) argued that 'brain science' as of that time (i.e. 1997), had little to offer in terms of educational practice, it was, in an oft-cited metaphor, a bridge too far. Bruer, amongst others, has more recently signed 'The Santiago Declaration' (2007), which states, in part: 'Current brain research offers a promissory note, however, for the future. Developmental models and our understanding of learning will be aided by studies that reveal the effects of experience on brain systems working in concert. This work is likely to enhance our understanding of the mechanisms underlying learning'. A key and potentially contentious issue here concerns 'the effects of experience'.

The term 'experience' and the effects thereof can be interpreted in many ways. From a materialist perspective, experience is what happens to the body, and the effects of that experience manifest as objective changes in body, brain, and behavior. From an idealist perspective, experience is what we have. That is to say, it is, quite literally, what we experience as subjective beings—and the effects thereof concern effects pertaining to our state of mind. It is possible, in both cases, to believe that one of these views or the other is simply an epiphenomenon or illusion, or that never the twain shall meet. In cases where some degree of interaction is admitted between mind and brain, there is an issue of priority concerning exactly what is causing what. Do subjective changes in experience cause objective changes in brain, or vice versa? There are many more sophisticated and subtle variants of the nature and relation between mind, brain, and the rest of the body than these. My aim is not to attempt a comprehensive treatment of these differences here, but rather, to point out that it is herein, in one's interpretation of this key phrase, 'the effects of experience', that one's philosophical commitments are most likely to be found.

Having said that, my own philosophical commitment in this regard, simply put, is to view mind, a.k.a. lived experience, as a bona fide property of matter. That is to say, mind is embodied in matter. It matters not to me if that property ultimately proves to be inherent or emergent, or a function of organization or an outcome of certain types of complex interaction, or whatever. What counts to me, as an educator, is that mind actually does have an effect on matter, and that means presupposing that one's mind can, at least to some extent, have causal effects on others, and on one's own brain and body in particular. *Prima facie*, such a view runs contrary to a fundamental philosophical commitment of most, if not all, scientists—viz., the notion that mind cannot have any causal efficacy whatsoever. To presuppose otherwise, however, would be to eliminate volition as a human characteristic, to render experience a matter of happenstance, and to deny any sense of moral agency or empowerment to learners. I find this unacceptable. Hence, I prefer to consider mind and brain different aspects of a unitary 'mindbrain', and to identify the study of the mindbrain as the true object of educational neuroscience.

Furthermore, as an educational neuroscientist, I view mind, following Varela, Thompson, and Rosch (1991), as embodied, but embodied in a radical sense such that any changes in subjective experience must in principle manifest objectively in some manner as changes in brain, body, and behavior, and vice versa. Placing the details of causality and the nature of matter aside, this radical view of the embodied mind (viz., the mindbrain) warrants a search for correlations between subjective experience and embodied behavior. If we wish to study subjectively experienced changes in the mental states of learners, one promising avenue in so doing is to study changes in brain and brain behavior. This is ostensibly the motivation for studies in psychophysiology and cognitive neuroscience. Accordingly, philosophical considerations aside, educational neuroscience can rightfully appropriate methods, theories, and share results with these disciplines.

The foregoing conception of educational neuroscience, in sum, seeks to bridge the gap between minds and bodies, with particular emphases on brains as our principal organs of thought, and thereby render the nature and various effects of educational experience more comprehensible and meaningful. Once again, I emphasize the plural here, as education involves the mindbrain in interaction with others, as well as with itself and various aspects of the world. Furthermore, although I have identified the mindbrain as the primary object of study for educational neuroscience, I view mind as manifest throughout the body and possibly extending beyond, but with the brain as its central and predominant locus.

In drawing on methods, theories, and results from cognitive neuroscience, educational neuroscience can be pursued as an *applied* cognitive neuroscience, but when considering philosophical commitments and methodological orientations (the latter about which I have more to say below), educational neuroscience has the potential to go beyond the prevalent mechanistic assumptions underlying those areas to become a truly transdisciplinary enterprise that also embraces radically empirical phenomenologies of lived experience. At the other end of this spectrum, research in educational neuroscience is also geared toward informing educational practice. As such it falls clearly within the purview of neuroeducation and more broadly conceived initiatives concerning mind, brain, and education, as evidenced in other contributions to this book.

Some Objections

To designate 'mindbrains' as the proper objects of educational neuroscience, then, is to emphasize that mind and brain are predominantly one thing, not two. For those who would doubt such a claim, I proffer the following points. First, in many jurisdictions, the absence of brain activity constitutes a legal definition of death. Second, it is clear, especially from electroencephalographic (EEG) studies, the bioelectromagnetic field generated by human brains clearly indicates subjectively experienced transitions between sleep and wakefulness. Third, EEG reveals profound alterations in brain behavior during seizures. Fourth, studies using transcranial magnetic stimulation (TMS) to alter brain behavior consistently and repeatedly alter mental experience. Fifth, lesion studies can dramatically illustrate effects thereof on mental functioning. Sixth, studies in brain plasticity indicate that living brains continually undergo massive changes in neural connectivity from moment to moment. Seventh, substances having noticeable effects on brain chemistry typically have noticeable effects on mental experience. Eighth, the burgeoning field of cognitive neuroscience, through a variety of brain imaging technologies, is providing mounting evidence that changes in cognitive function are significantly correlated with changes in brain and brain behavior; for instance, volitional alterations in mental state associated with meditation, attention, reasoning and recall. Even a cursory search of the burgeoning literature in cognitive neuroscience, and more recently and incipiently, neurophenomenology (inspired by Francisco Varela and colleagues), reveals growing insight into correlations between mind and brain with no signs of abatement in sight.

Returning, now, to my opening question: What does the brain have to do with learning? This question is often on the minds of educational practitioners (brain-based enthusiasts notwithstanding). After all, it is quite clear that kids in classrooms are a far cry from neurons in Petri dishes. Moreover, what need is there for teachers concerned with cultivating intellectual and moral development to concern themselves with the complexities and mechanisms of brain anatomy and physiology? These are legitimate concerns, but they miss the point. The question is not whether there are connections between minds and brains. There clearly are. The evidence is insurmountable and growing. The question, then, is to what extent, subject to intrinsic theoretical and practical limits of measurement and analysis, can we identify changes in mental states as changes in brain and brain behavior, and vice versa—insofar as such changes in, and the development of the mindbrain, affect and concern educational matters associated with teaching and learning. Such is the charge and mandate of educational neuroscience. How, then, to best go about this?

Motivations, Aims, and Prospects

Motivations are many and varied things. There are some cognitive neuroscientists who are interested in studying and contributing to educational problems associated with teaching and learning. To the extent that they bring their expertise to bear in these ways, educational neuroscience can be seen as an applied cognitive neuroscience. On the other hand, there are also some educational researchers, practitioners, and policy makers who are interested in understanding ways in which findings from the neurosciences may be relevant to improving education. To the extent that they bring their expertise to bear in these ways, educational neuroscience can be seen as part of neuroeducation and the growing initiatives concerning relations between mind, brain, and education.

As both a philosopher of education and researcher in mathematics education (Campbell, 1998), I see educational neuroscience as presenting unique opportunities to tackle fundamental problems in both of these areas simultaneously and in a coordinated fashion. In the foregoing sections I have focused on philosophical issues germane to defining educational neuroscience as a bona fide transdisciplinary area of educational research—a new area of educational research primarily concerned with the nature and educational development of the mindbrain. Of what possible relevance might this be to research in mathematics education?

There are many problems in mathematics education research, but those I am most interested in, and consider as fundamental, are the nature of mathematics anxiety and mathematical concept formation. I am particularly interested in ways in which the former impedes the latter, and ways in which the latter can mitigate the former. My chosen target population is preservice teachers, as I believe it is this population that ultimately has the greatest influence toward improving the effectiveness of mathematics education (see e.g. Campbell & Zazkis, 2002; Zazkis & Campbell, 2006).

Learner responses to mathematics anxiety and mathematical illumination typically range from poignant to salient. Both constitute deeply embodied phenomena, and thereby present themselves as natural candidates for research in educational neuroscience. There are other issues of pressing concern in mathematics education that make for good topics of study using methods of educational neuroscience, such as problem solving and instructional design, especially involving computer-user interfacing with computer-based educational software. I think these kinds of educational problems are particularly well suited for study from a radically embodied perspective.

Theories, Methods, and Collaborations

Embodied cognition affords educational neuroscience a firm methodological foundation as well as a strong philosophical framework for the study of such problems, and it has been my aim to contribute to the development of both with that end in view. I mentioned earlier that educational neuroscientists could conduct studies using a variety of research paradigms whilst being motivated by and drawing conclusions based upon differences in philosophical commitments (Campbell, forthcoming). The so-called 'paradigm wars' that have beset educational research over the past few decades have in my view been based more upon fundamental philosophical differences than on the much more specific theories and methods associated with various research paradigms. As a transdisciplinary enterprise within a framework of embodied cognition, I see no reason why educational neuroscience should be restricted when it comes to research paradigms.

For instance, qualitative research in educational research often strives to understand a given problem in depth as it is situated. The aim is to gain deep understanding in order to provide some insight into a given condition, and/or how it arose or could be improved. Sometimes, new and often unanticipated factors and phenomena are revealed. There is less consideration given as to whether qualitative findings have more general application, as such questions more rightfully fall under the purview of quantitative research. The theories, methods, instruments, and aims of qualitative and quantitative research differ, but there is no overarching reason to see them as inherently incommensurable, unless, that is, for some reason one or the other is at odds with one's philosophical commitments. However, with the growing prevalence of pragmatism and emphasis on mixed methods should come greater openness toward novel approaches to educational research enabled through educational neuroscience.

From an embodied perspective, it is just as important to understand how cognition is situated within individuals and classrooms, how learning comes about and can be enhanced, as it is to gain knowledge of brain mechanisms underlying cognitive function. Hence, both cognitive and *social* neuroscience have much to offer educational neuroscience. In applying results of cognitive and social neuroscience to educational problems, it behooves educational neuroscientists to work closely, either in direct collaboration with, or through assimilating relevant literature generated by, these neuroscientists. Not only do the neurosciences have much to offer in terms of results, but also in terms of theories and methods new to educational research, especially with regard to new technologies in brain imaging (Campbell, 2010; Makeig, 2010).

In my empirical research in mathematics education, coupled with a long-standing philosophical commitment to embodied cognition (e.g. Campbell, 2002; Campbell & Dawson, 1995), I have endeavored to bring to bear as much observational control of the learning process as possible (Campbell, 2003; Campbell, with the ENL Group, 2007). Currently in my educational neuroscience laboratory (a.k.a. the ENGRAMMETRON), in the Faculty of Education at Simon Fraser University, I have equipment to record electroencephalograms (EEG), electrocardiograms (EKG), electrooculograms (EOG), and electromylograms (EMG), which pertain to brain activity, heart rate, eye movement, and muscle movement, respectively. I can also record respiration, blood flow volume, and skin conductance. All these psychophysiological metrics are augmented with eye-tracking technology, screen capture, keyboard and mouse capture, and multiple video recordings of participants from various perspectives. These data sets can then be integrated and synchronized for coding, analysis, and interpretation, thereby affording comprehensive observations and insights into the learning process (Campbell, 2010).

Beyond facilitating my own research, I conceived, designed, and implemented the ENGRAMMETRON <http://www.engrammetron.net> 'from the get go' as a state-of-the-art facility for collaboration, not only in educational neuroscience, but also for bringing greater observational control to empirical research in education in general. As such, the ENGRAMMETRON serves as a hub for a strategic research cluster of researchers in education, psychology, cognitive neuroscience, biomedical engineering, and so on, called ENGRAM/ME (Educational Neuroscience Group for Research on Affect and Mentation / in Mathematics Education), resulting thus far in a variety of completed, on-going, and anticipated collaborations.

Such collaborations with students and ENGRAM/ME members have been helpful in bringing diverse sets of expertise to bear on a project-by-project basis (e.g. Bigdeli & Campbell, 2007; Campbell & Handscomb, 2007; Campbell, Cimen & Handscomb, 2009; Campbell, Handscomb, Zaparyniuk, Sha, Cimen & Shipulina, 2009; Campbell, Siddo & Bigdeli, 2007; de Castell, Campbell, Jenson, Shipulina, Cimen & Taylor, 2010; Du, Campbell & Kaufman, 2010; Patten & Campbell, 2007; Sha, Winne & Campbell, 2009; Shipulina, Campbell & Cimen, 2009), with other projects planned and in progress.

Challenges, Results, and Implications

Thus far, the emphasis in this initiative in educational neuroscience has been primarily focused on identifying, meeting, and overcoming challenges. The main challenge has been to muster evidence and rationale to justify this initiative to funding agencies traditionally supporting educational research (Campbell, 2005). The *Canada Foundation for Innovation*, the *British Columbia Knowledge Development Fund*, and Simon Fraser University have collectively funded the development of my lab, and the *Social Science and Humanities Research Council of Canada* has supported initial efforts for conducting research therein. Another huge challenge has been to transform the lab from an idea to a reality. The task was daunting, but there are benefits from being able to design such a facility from the ground up (Campbell, 2010). In essence, and what makes the ENGRAMMETRON such a unique facility, is that it integrates observational methods and techniques from cognitive neuroscience, psychophysiology, cognitive psychology, with audiovisual recordings prevalent in contemporary empirical educational research.

The main result in meeting these two challenges of funding and building the lab has been to provide a leading edge facility that augments traditional empirical educational research in such a manner that educational researchers can design experiments that also investigate and apply results pertaining to brain, body, and behavior obtained from the neurosciences and psychophysiology in a manner that heretofore has not been possible. Despite ample progress made to date, there remain on-going challenges in refining and applying our philosophical framework, and various theories, methods and techniques associated with acquiring and analyzing such a diverse set of observations—but these are topics for which there is neither need nor sufficient space to discuss here (see, for instance, e.g. Campbell, Cimen & Handscomb, 2009; Shipulina *et al.*, 2009).

There is also an on-going challenge of attracting and training highly qualified personnel. My initial impulse in this regard was to take a 'build it and they will come' attitude, and to some extent, thus far, I am happy to report that this has come to pass, and hopefully will continue to do so the more word gets out regarding this facility. On a more proactive basis, I have instituted a graduate level course that I have offered three times now with substantial success (Campbell, 2008).

Despite these on-going challenges coupled with, as yet in these early days of operation, relatively modest but promising preliminary results, the implications are clear. Empirical research in education must move full bore into the 21st century. Perhaps never in the history of our species has it been more important to cultivate our individual and collective understanding of our selves, each other, and this finite world that we inhabit. From a scientific perspective, as well as an embodied perspective, the more observational perspectives that we can muster in educational research, the better opportunities we will have in measuring, identifying, and comprehending new phenomena and significant factors associated with cognitive and social development germane to various aspects of teaching and learning. In much the same way audiovisual recordings augmented field notes, so too can methods of educational neuroscience augment audiovisual recordings. Affording new avenues for experimental design and collaboration, be it pursued in a multidisciplinary, interdisciplinary, or transdisciplinary manner, educational neuroscience has significant potential to inform educational philosophy and theory, and thereby help to both elevate and edify our understanding of the human condition.

References

Bruer, J. T. (1997) Education and the Brain: A bridge too far, *Educational Researcher*, 26:8, pp. 4–16.

Bigdeli, S. & Campbell, S. R. (2007) Psychophysiology of ESL anxiety: New possibilities. Paper presented to the Canadian Educational Researchers' Association at the Annual Canadian Society for Studies in Education Conference, May (Saskatoon, SK).

Campbell, S. R. (1998) *Preservice Teachers' Understanding of Elementary Number Theory: Qualitative constructivist research situated within a Kantian framework for understanding educational inquiry* (Burnaby, BC, Simon Fraser University).

Campbell, S. R. (2002) Constructivism and the Limits of Reason: Revisiting the Kantian problematic, *Studies in Philosophy and Education*, 21:6, pp. 421–445.

Campbell, S. R. (2003) Dynamic Tracking of Preservice Teachers' Experiences with Computer-Based Mathematics Learning Environments, *Mathematics Education Research Journal*, 15:1, pp. 70–82.

Campbell, S. R. (2004) Forward and Inverse Modelling: Mathematics at the nexus between mind and world, in: H-W. Henn & W. Blum (eds), *ICMI Study 14: Applications and modelling in mathematics education—pre-conference volume* (Dortmund, University of Dortmund), pp. 59–64.

Campbell, S. R. (2005) Specification and Rationale for Establishing a Mathematics Educational Neuroscience Laboratory. Paper presented to the meeting of the American Educational Research Association (Montreal, QC).

Campbell, S. R. (2006a) Educational Neuroscience: New horizons for research in mathematics education, in: J. Novotná, H. Moraová, M. Krátká & N. Stehlíková (eds), *Proceedings 30 Conference of the International Group for the Psychology of Mathematics Education*, Vol. 2 (Prague, PME), pp. 257–264.

Campbell, S. R. (2006b) Defining Mathematics Educational Neuroscience, in: S. Alatorre, J. L. Cortina, M. Sáiz & A. Méndez (eds), *Proceedings of the 28th Annual Meeting of the North American Chapter of the International Group for the Psychology of Mathematics Education*, Vol. 2 (Mérida, Universidad Pedagógica Nacional), pp. 442–449.

Campbell, S. R., with the ENL Group (2007) The ENGRAMMETRON: Establishing an educational neuroscience laboratory, *Simon Fraser University Educational Review*, 1, pp. 17–29.

Campbell, S. R. (2008) Launching a Graduate Course in Educational Neuroscience, *Simon Fraser University Educational Review*, 2, pp. 39–51.

Campbell, S. R. (2010) Embodied Minds and Dancing Brains: New opportunities for research in mathematics education, in: B. Sriraman & L. English (eds), *Theories of Mathematics Education: Seeking new frontiers* (Berlin, Springer), pp. 309–331.

Campbell, S. R. (forthcoming) Philosophical Frameworks and Research Paradigms in Educational Theory and Research, *Journal of Philosophy of Education*.

Campbell, S. R. & Dawson, A. J. (1995) Learning as Embodied Action, in: R. Sutherland & J. Mason (eds), *NATO Advanced Research Workshop: Exploiting mental imagery with computers in mathematics education*, NATO ASI Series F, vol. 138 (Berlin, Springer), pp. 233–249.

Campbell, S. R., Cimen, O. A. & Handscomb, K. (2009) Learning and Understanding Division: A study in educational neuroscience. Paper presented to the American Educational Research Association: Brain, Neuroscience, and Education SIG, April (San Diego, CA) (ED505739).

Campbell, S. R. & Handscomb, K. (2007) An Embodied View of Mind-Body Correlates. Paper presented to the American Educational Research Association: Brain, Neuroscience, and Education SIG, April (Chicago, IL).

Campbell, S. R., Handscomb, K., Zaparyniuk, N. E., Sha, L., Cimen, O. A. & Shipulina, O. V. (2009) Investigating Image-Based Perception and Reasoning in Geometry. Paper presented to the American Educational Research Association: Brain, Neuroscience, and Education SIG, April (San Diego, CA) (ED505740).

Campbell, S. R., Siddo, R. A. & Bigdeli, S. (2007) Integrating Psychometrics and Psychophysiology in the Study of Elementary Preservice Teachers' Anxieties Toward Teaching and Learning Mathematics. Paper presented to the American Educational Research Association: Brain, Neuroscience, and Education SIG, April (Chicago, IL).

Campbell, S. R. & Zazkis, R. (eds) (2002) *Learning and Teaching Number Theory: Research in cognition and instruction* (Westport, CT, Ablex Publishing).

Chalmers, D. J. (1995) Facing Up to the Problem of Consciousness, *Journal of Consciousness Studies*, 2:3, pp. 200–219.

De Castell, S., Campbell, S. R., Jenson, J., Shipulina, O. V., Cimen, O. A. & Taylor, N. (2010) The Eyes Have It: Gender difference and spatial orientation in video games. Paper presented to CHI 2010 (28th Annual ACM Conference on Human Factors in Computing Systems) Workshop on Video Games as Research Instruments, April (Atlanta, GA).

Du, X., Campbell, S. R. & Kaufman, D. (2010) A Study of Biofeedback in a Gaming Environment, in: D. Kaufman & L. Suavé (eds), *Educational Gameplay and Simulation Environments: Case studies and lessons learned* (Hershey, PA, IGI Global), pp. 326–345.

Gibbons, M., Limoges, C., Nowotny, H., Schwartzman, S., Scott, P. & Trow, M. (1994) *The New Production of Knowledge* (London, Sage).

Makeig, S. (2010) Commentary on Embodied Minds and Dancing Brains: New opportunities for research in mathematics education, in: B. Sriraman and L. English (eds), *Theories of Mathematics Education: Seeking new frontiers* (Berlin, Springer), pp. 333–337.

Patten, K. & Campbell, S. R. (2007) *From Educational Neuroscience to Neuropedagogy*. Paper presented to the Canadian Association for Educational Psychology at the Annual Canadian Society for Studies in Education Conference, May (Saskatoon, SK).

Santiago Declaration (2007) Available online at <www.jsmf.org/santiagodeclaration/> (Accessed 31 October 2009).

Sha, L., Winne, P. H. & Campbell, S. R. (2009) Personal Factors Underlying the Relation Between Metacognitive Judgments and Control in Self-Regulated Learning. Paper presented to the American Educational Research Association 2009 Annual Meeting, April (San Diego, CA).

Shipulina, O. V., Campbell, S. R. & Cimen, O. A. (2009) Electrooculography: Connecting mind, brain, and behavior in mathematics education research. Paper presented to the Brain, Neuroscience, and Education SIG at 2009 AERA Annual Meeting, April (San Diego, CA) (ED505692).

Tokuhama-Espinosa, T. N. (2008) The Scientifically Substantiated Art of Teaching: A study in the development of standards in the new academic field of neuroeducation (mind, brain, and education science). PhD thesis, Capella University, Minneapolis, MN.

Varela, F. J., Thompson, E. & Rosch, E. (1991) *The Embodied Mind: Cognitive science and human experience* (Cambridge, MA, MIT Press).

Velmans, M. (1995) The Relation of Consciousness to the Material World, *Journal of Consciousness Studies*, 2:3, pp. 255–265.

Zazkis, R. & Campbell, S. R. (eds) (2006) *Number Theory in Mathematics Education: Perspectives and prospects* (Mahwah, NJ, Lawrence Erlbaum Associates).

3

Can Cognitive Neuroscience Ground a Science of Learning?

ANTHONY E. KELLY

Why is there a current interest in cognitive neuroscience findings? In spite of the pessimism of Bruer (1997) and the more recent caveats of Varma, McCandliss, and Schwartz (2008), and Willingham (2008), the past decade has seen an upsurge in studies focusing on the brain-basis for learning (see OECD, 2007 for a comprehensive review of brain-related research in education). The following factors appear to be contributing to this interest:

- A desire to scientifically debunk popular 'brain-based' claims about learning and teaching (i.e. 'neuromythologies');
- A growing set of studies on the neural bases for mathematical thought;
- The establishment of recent gains in understanding the brain bases for processes of decoding in reading;
- Decades of behavioral and cognitive science findings on both reading and learning mathematics upon which to base brain studies in these areas;
- Frustration with vague theories of learning and the desire to disambiguate and constrain research hypotheses at the behavioral, cognitive and social levels of analysis;
- Frustration with broad measures of achievement (often paper-and-pencil standardized tests) that do not allow the ability to sharpen and ground diagnosis and remediation of learning difficulties;
- A desire to introduce and explore new mixed-methods research methodologies in the social sciences;
- A sense of urgency to address emerging ethical issues that pertain both to neuroscience and to learning;
- The ongoing goal to improve methods of teaching worldwide, including the quality of educational materials;
- The emergence of more comprehensive and testable models of learning emerging from cognitive science that can bridge learning and cognitive psychology;
- A desire to understand and promote creativity, and to explore cognition in music and other areas;
- The challenges neuroscientists face in modeling learning phenomena continue to push the boundaries of imaging technologies, and of the expertise required to co-formulate clinical learning tasks with learning scientists.

Educational Neuroscience, First Edition. Edited by Kathryn E. Patten and Stephen R. Campbell.
Chapters © 2011 The Authors. Book compilation © 2011 Educational Philosophy and Theory/Blackwell Publishing Ltd.
Published 2011 by Blackwell Publishing Ltd.

Part of the difficulty that educational psychology has faced in the study of learning is that, too often, learning constructs exist as psychometric objects measured by tests (cf. the treatment of intelligence, Sternberg, 2007). At the behavioral level, particularly in ethnographic studies of learning in classrooms, researchers typically lack the knowledge to differentiate between activity that is essentially accidental or contingent (i.e. behavior A happened as a result of behavior B, but the results are transitory and the implications for understanding are fleeting) and behaviors that point to more fundamental processes or constraints that are necessary to a scientific understanding (e.g. Kelly, 2004; Kelly, 2008). I argue that a science of learning can be grounded in a set of empirical primitives and that these primitives are becoming known via cognitive neuroscience-based analyses.

I recognize that the state of the art in cognitive neuroscience research is still in its early stages (OECD, 2007; Frith & Blakemore, 2005), and that cognitive neuroscience will take decades to mature. Nonetheless, important questions are being addressed, and new ones will be answered, by advances in functional imaging technology such as near infrared spectroscopy (Koh *et al.*, 2007).

Moreover, there already is evidence that general abilities that define learning, such as studies of general understanding of concrete and abstract concepts, are linked to specific brain systems (Binder *et al.*, 2005). Other studies have focused on attention (e.g. Rueda *et al.*, 2004), and executive function, attention and memory (Fan *et al.*, 2003; Fan *et al.*, 2005; Fossella *et al.*, 2002). Interestingly, the attention capacity may have genetic triggers (Parasuraman *et al.*, 2005).

New work is appearing on long-term memory consolidation as a complex neurobiological process, involving synaptogenesis and neurogenesis (Shors, 2008). For related studies on memory, see Kesner, 2009; Reder, Park & Kieffaber, 2009; and Weinberger *et al.*, 2009.

Specific abilities, such as mathematics, are also apparently rooted in and require brain circuitry to support simple function and these circuits also underlie significant performance deficits. For example, mathematical ability not only appears quite early in life (e.g. Berger, Tzur & Posner, 2006; Wynn, 1992), but is also being found to have phylogenetic roots and may not be 'uniquely human' (e.g. Diester & Nieder, 2007; Nieder, Diester & Tudusciuc, 2006).

When mathematical ability is impaired in humans, growing evidence points to a brain basis, especially in extreme cases such as in dyscalculia (e.g. Butterworth, 2005), or in cases of dementia, cognitive decline appears to be related to specific dissociations in numerical ability (Cappelletti *et al.*, 2005, see also, Tang, Ward & Butterworth, 2008). On the positive side, intensive instruction in multiplication and subtraction appears to impact different neural circuitry (Ischebeck *et al.*, 2006). In fact, Dehaene and colleagues argue for a brain basis for 'number sense', generally (Dehaene, 1997; Feigenson, Dehaene & Spelke, 2004; Hubbard, Piazza, Pinel & Dehaene, 2005), which they contend is the result of evolutionary pressures.

The brain circuitry for reading is already well established, building on decades of behavioral work. Like mathematics research, there is growing evidence of very early circuitry development to support reading skill (e.g. Guttorm *et al.*, 2005; Molfese, 2000; Schlaggar & McCandliss, 2007). For example, there are studies implicating the posterior cortex in development dyslexia (e.g. Pugh, Mencl, Shaywitz *et al.*, 2000; Pugh, Mencl,

Jenner *et al.*, 2000; Pugh *et al.*, 2005; Shaywitz *et al.*, 2002). It is important to note that neural studies go beyond simply mapping areas of the brain to behavioral activity. Studies are now appearing that attempt to design learning based on brain studies such as executive attention (Rueda *et al.*, 2005), reading (e.g. Eden & Moats, 2002; McCandliss *et al.*, 2003; Sarkari *et al.*, 2002), and number sense (e.g. Wilson, Dehaene *et al.*, 2006; Wilson, Revkin *et al.*, 2006). Dehaene (2009) is a superb review of the neural basis for reading.

Significantly, studies are now appearing linking learning to changes in the brain via training interventions, completing an observational-correlational-experimental loop (e.g. Sandak *et al.*, 2004; Simos *et al.*, 2002; Temple *et al.*, 2003). The drivers in these studies are the tasks, which can be simultaneously tested within the limitations of current imaging technology, yet are informative about model building in learning (see Dehaene, 2008, for pointers on task design in arithmetic). New neuroscience studies are also emerging that link learning to more traditional factors such as socioeconomic conditions (Noble, McCandliss & Farah, 2007).

Most importantly, funding agencies are beginning to support research at the intersection of brain-based studies and learning. In 2008, the Research and Evaluation on Education in Science and Engineering (REESE) program at the US National Science Foundation targeted studies on the neural basis for mathematics learning. In 2009, the REESE call for proposals has been extended to all science, technology, engineering, and mathematics (STEM) learning:

1. Neural basis of STEM learning

Fundamental aspects of STEM learning are beginning to be understood in terms of neural processes and biological context. Discoveries in these and other areas are influencing our understanding of behavior, cognition, and the nature of human learning. REESE will support studies focused on human learning in the STEM fields drawing on a wide range of theoretical approaches and empirical techniques. It is incumbent upon those submitting proposals to make explicit the implications their work has for current theories of learning and instructional methods, however long-term and indirect they may be. For example, neuroscientific studies of attention or inhibition could constrain theories about the learning of specific STEM content or help explain why some misconceptions are robust and difficult to overcome. They could similarly inform the creation of principles of design for the development of instructional materials, informal learning opportunities, or the education of teachers in the STEM fields.

In order to gain traction on fundamental questions of mind and brain as related to STEM learning, REESE supports innovative combinations of theory, methods, and levels of analysis from a wide range of disciplines. An important aspect of these activities is to build capacity in neuroscience related to complex human learning and education, and to identify trajectories by which multidisciplinary research anchored in the biological basis of human learning can inform STEM educational practice. The involvement of researchers familiar with STEM educational practice will be of benefit both in helping to set the cognitive and neuroscientific research agendas in learning as well as in helping to disseminate relevant literatures across disciplines. (REESE Program Description. Available online at: http://nsf.gov/pubs/2008/nsf08585/nsf08585.htm)

Taken together, I contend that we are beginning to see, across these factors, the basis for a revolution in theorizing about learning that designs and refines its measures, guides its hypotheses, informs its analyses and grounds its conclusions using data from cognitive neuroscience studies. I expect that current, apparently incommensurate, theories or general descriptions about learning will be decided more and more on the basis of this growing empirical record. Theories are never abandoned easily, of course, but the disambiguation of claims at the hypothesis testing level using cognitive neuroscience data is likely to place upward pressure on theories, which are too often contingent descriptions of learning with little specification of mechanism or grounding in the larger set of findings in science.

References

Berger, A., Tzur, G. & Posner, M. I. (2006) Infant Brains Detect Arithmetic Errors, *Proceedings of the National Academy of Sciences*, 103:33, pp. 12649–12653.

Binder, J. R., Westbury, C. F., McKiernan, K. A., Possing E. T. & Medler, D. A. (2005) Distinct Brain Systems for Processing Concrete and Abstract Concepts, *Journal of Cognitive Neuroscience*, 17, pp. 905–917.

Bruer, J. T. (1997) Education and the Brain: A bridge too far, *Educational Researcher*, 26, pp. 4–16.

Butterworth, B. (2005) Developmental Dyscalculia, in: J. Campbell (ed.), *Handbook of Mathematical Cognition* (New York, Psychology Press).

Cappelletti, M., Kopelman, M. D., Morton, J. & Butterworth, B. (2005) Dissociations in Numerical Abilities Revealed by Progressive Cognitive Decline in a Patient with Semantic Dementia, *Cognitive Neuropsychology*, 22:7, pp. 771–793.

Dehaene, S. (1997) *The Number Sense: How the mind creates mathematics* (New York, Oxford University Press).

Dehaene, S. (2008) Can Cognitive Neuroscience Help Design Innovative Education Protocols? The case of arithmetic, in: Académie des Sciences (ed.), *Education, sciences cognitives et neurosciences* (Paris, Presses Universitaires de France), pp. 41–48.

Dehaene, S. (2009) *Reading in the Brain: The science and evolution of a human invention* (New York, Viking).

Diester, I. & Nieder. A. (2007) Semantic Associations between Signs and Numerical Categories in the Prefrontal Cortex, *PLoS Biology*, 5:11, e294 doi:10.1371/journal.pbio.0050294.

Eden. G. & Moats, L. (2002) The Role of Neuroscience in the Remediation of Students with Dyslexia, *Nature Neuroscience* Supplement, 5, pp. 1080–1084.

Fan, J., Fossella, J. A., Summer T. & Posner, M. I. (2003) Mapping the Genetic Variation of Executive Attention onto Brain Activity, *Proceedings of the National Academy of Sciences of the USA*, 100, pp. 7406–7411.

Fan, J., McCandliss, B. D., Fossella, J., Flombaum, J. I. & Posner, M. I. (2005) The Activation of Attentional Networks, *Neuroimage*, 26, pp. 471–479.

Feigenson, L., Dehaene, S. & Spelke. E. (2004) Core Systems of Number, *Trends in Cognitive Sciences*, 8:7, pp. 307–314.

Fossella, J., Sommer T., Fan, J., Wu, Y., Swanson, J. M., Pfaff, D. W. & Posner, M. I. (2002) Assessing the Molecular Genetics of Attention Networks, *BMC Neuroscience*, 3:14, doi:10.1186/1471-2202-3-14.

Frith, U. & Blakemore, S-J. (2005) *The Learning Brain: Lessons for education* (Malden, MA, Blackwell).

Guttorm, T. K., Leppanen, P. H. T., Poikkeus, A. M., Eklund, K. M., Lyytinen, P. & Lyytinen, H. (2005) Brain Event-related Potentials (ERPs) Measured at Birth Predict Later Language Development in Children with and without Familial Risk for Dyslexia, *Cortex*, 41:3, pp. 291–303.

Hubbard, E. M., Piazza, M., Pinel, P. & Dehaene, S. (2005) Interactions between Number and Space in Parietal Cortex, *Nature Reviews Neuroscience*, 6, pp. 435–448.

Ischebeck, A., Zamarian, L., Siedentopf, C., Koppelstatter, F., Benke, T., Felber, S. & Delazer, M. (2006) How Specifically Do We Learn? Imaging the learning of multiplication and subtraction, *NeuroImage*, 30, pp. 1365–1375.

Kelly, A. E. (2004) Design Research in Education: Yes, but is it methodological, *Journal of the Learning Sciences*, 13:1, pp. 115–128.

Kelly, A. E. (2008) Brain Research and Education: Potential implications for pedagogy, in: Académie des Sciences (ed.), *Education, sciences cognitives et neurosciences* (Paris, Presses Universitaires de France), pp. 75–91.

Kesner, R. (2009) Tapestry of Memory, *Behavioral Neuroscience*, 123, pp. 1–13.

Koh, P. H., Glaser, D. E., Flandin, G., Kiebel, S., Butterworth, B., Maki, A., Delpy, D. T. & Elwell, C. E. (2007) Functional Optical Signal Analysis: A software tool for near-infrared spectroscopy data processing incorporating statistical parametric mapping, *Journal of Biomedical Optics*, 12:6, 064010-064011-064010-064013.

McCandliss, B., Beck, I.L., Sandak, R. & Perfetti, C. (2003) Focusing Attention on Decoding for Children with Poor Reading Skill: Design and preliminary test of the word building intervention, *Scientific Studies of Reading*, 7, pp. 75–104.

Molfese, D. L. (2000) Predicting Dyslexia at 8 Years of Age Using Neonatal Brain Responses, *Brain and Language*, 72:3, pp. 238–245.

Nieder, A., Diester L. & Tudusciuc, O. (2006) Temporal and Spatial Enumeration Processes in the Primate Parietal Cortex, *Science*, 313:5792, pp. 1431–1435.

Noble, K. G., McCandliss, B. D. & Farah, M. (2007) Socioeconomic Gradients Predict Individual Differences in Neurocognitive Abilities, *Developmental Science*, 10, pp. 464–480.

OECD (2007) *Understanding the Brain: The birth of a new learning science* (Paris, OECD Publishing).

Parasuraman, R., Greenwood, P.M., Kumar, R. & Fossella, J. (2005) Beyond Heritability: Neurotransmitter genes differentially modulate visuospatial attention and working memory, *Psychological Science*, 16, pp. 200–207.

Pugh, K., Mencl, E. W., Shaywitz, B. A., Shaywitz, S. E., Fulbright, R. K., Skudlarski, P., Constable, R. T., Marchione, K., Jenner A.R., Shankweiler, D. P., Katz, L., Fletcher, J., Lacadie, C. & Gore, J. C. (2000) The Angular Gyrus in Developmental Dyslexia: Task-specific differences in functional connectivity in posterior cortex, *Psychological Science*, 11, pp. 51–56.

Pugh, K. R., Mencl, W. E., Jenner, A. J., Katz, L., Frost, S. J., Lee, J. R., Shaywitz, S. E. & Shaywitz, B. A. (2000) Functional Neuroimaging Studies of Reading and Reading Disability (Developmental Dyslexia), *Mental Retardation and Developmental Disabilites Review*, 6:3, pp. 207–213.

Pugh, K. R., Sandak, R., Frost, S. J., Moore, D. & Mencl, W. E. (2005) Examining Reading Development and Reading Disability in English Language Learners: Potential contributions from functional neuroimaging, *Learning Disabilities Research & Practice*, 20:1, pp. 24–30.

Reder L. M., Park, H. & Kieffaber, P. D. (2009) Memory Systems Do Not Divide on Consciousness: Reinterpreting memory in terms of activation and binding, *Psychological Bulletin*, 135:1, pp. 23–49.

Rueda, M. R., Fan, J., Halparin, J., Gruber, D., Lercari, L. P., McCandliss, BD. & Posner, MI. (2004) Development of Attention During Childhood, *Neuropsychologia*, 42, pp. 1029–1040.

Rueda, M. R., Rothbart, M. K., Saccamanno, L. & Posner, M. I. (2005) Training, Maturation and Genetic Influences on the Development of Executive Attention, *Proceedings of the National Academy of Sciences of the USA*, 102, pp. 14931–14936.

Sandak, R., Mencl, W. E., Frost, S. J., Mason, S. A., Rueckl, J. G., Katz, L., Moore, D. L., Mason, S. A., Fulbright, R., Constable, R. T. & Pugh, K. R. (2004) The Neurobiology of Adaptive Learning in Reading: A contrast of different training conditions, *Cognitive Affective and Behavioral Neuroscience*, 4, pp. 67–88.

Sarkari, S., Simos, P. G., Fletcher, J. M., Castillo, E. M., Breier, J. I. & Papanicolaou, A. C. (2002) The Emergence and Treatment of Developmental Reading Disability: Contributions of functional brain imaging, *Seminars in Pediatric Neurology*, 9, pp. 227–236.

Schlaggar, B. L. & McCandliss, B. D. (2007) Development of Neural Systems for Reading, *Annual Review of Neuroscience*, 30, pp. 475–503.

Shaywitz, B. A., Shaywitz, S. E., Pugh, K. R., Mencl, W. E., Fulbright, R. K., Skudlarski, P., Constable, R. T., Marchione, K. E., Fletcher, J. M., Lyon, G. R. & Gore, J. C. (2002) Disruption of Posterior Brain Systems for Reading in Children with Developmental Dyslexia, *Biological Psychiatry*, 52, pp. 101–110.

Shors, T. J. (2008) From Stem Cells to Grandmother Cells: How neurogenesis relates to learning and memory, *Cell Stem Cell*, 3, pp. 253–258.

Simos, P. G., Fletcher, J. M., Bergman, E., Breier, J. I., Foorman, B. R., Castillo, E. M., Davis, R. N., Fitzgerald, M. & Papanicolaou, A. C. (2002) Dyslexia-specific Brain Activation Profile Becomes Normal Following Successful Remedial Training, *Neurology*, 58, pp. 1203–1213.

Sternberg, R. (ed.) (2007) *International Handbook of Intelligence* (Cambridge, Cambridge University Press).

Tang, J., Ward, J. & Butterworth, B. (2008) Number Forms in the Brain, *Journal of Cognitive Neuroscience*, 20:9, pp. 1547–1556.

Temple, E., Deutsch, G. K., Poldrack, R. A., Miller, S. L., Tallal, P., Merzenich, M. M. & Gabrieli, J. D. E. (2003) Neural Deficits in Children with Dyslexia Ameliorated by Behavioral Remediation: Evidence from functional MRI, *Proceedings of the National Academy of Sciences*, 100, pp. 2860–2865.

Varma, S., McCandliss, B. D. & Schwartz, D. L. (2008) Scientific and Pragmatic Challenges for Bridging Education and Neuroscience, *Educational Researcher*, 37, pp. 140–152.

Weinberger, N. M., Miasnikov, A. A. & Chen, J. C. (2009) Sensory Memory Consolidation Observed: Increased specificity of detail over days, *Neurobiology of Learning and Memory*, 91, pp. 273–286.

Willingham, D. (2008) When and How Neuroscience Applies to Education, *Phi Delta Kappan*, 89, pp. 421–423.

Wilson, A. J., Dehaene, S., Pinel, P., Revkin, S. K., Cohen, L. & Cohen, D. (2006) Principles Underlying the Design of 'The Number Race', an Adaptive Computer Game for Remediation of Dyscalculia, *Behavioral and Brain Functions*, 2:19. Available online at: http://www.behavioralandbrainfunctions.com/content/2/1/19

Wilson, A. J., Revkin, S. K., Cohen, D., Cohen, L. & Dehaene, S. (2006) An Open Trial Assessment of 'The Number Race', an Adaptive Computer Game for Remediation of Dyscalculia. *Behavioral and Brain Functions*, 2:20. Available online at: http://www.unicog.org/publications/WilsonDehaene_2006_BehBrainFunctions_OpenTrialAssessment.pdf

Wynn, K. (1992) Addition and Subtraction by Human Infants, *Nature*, 358, pp. 749–750.

4

A Multiperspective Approach to Neuroeducational Research

Paul A. Howard-Jones

Neuroeducational[1] Research and the Interrelation of Diverse Perspectives on Learning

To include concepts of brain function in educational thinking appears a common sense notion that has become popular with many educators (Pickering & Howard-Jones, 2007) and is stimulating discussion internationally, as evidenced by the recent OECD Brain and Learning project (OECD, 2007). In the UK, the NeuroEducational research network (NEnet, www.neuroeducational.net) at the University of Bristol has played a key role in recent national efforts to develop collaboration between the fields of neuroscience and education. In 2005–2006, NEnet co-ordinated an interdisciplinary seminar series, funded by the UK research councils, entitled 'Collaborative Frameworks in Neuroscience and Education'. Over 400 teachers, neuroscientists, educational psychologists, researchers and policy-makers met over six events across the country to discuss the issues and opportunities that might be provided by such a venture. The series gave rise to a commentary, whose popularity (downloading 110,000 copies in the first 6 months) demonstrated the rapidly growing and broadly-based educational interest in the brain (Howard-Jones, 2007). The commentary emphasized the need for two-way dialogue and for projects in which neuroscience and education collaborate in terms of both fundamental research and in the communication of its concepts. Such a two-way approach to ventures in neuroscience and education can serve two principle aims. The first aim is to enrich, develop and promulgate educational understanding and practice (Geake, 2004). The second aim, interrelating with the first, is to further scientific understanding of behaviours associated with learning, through the study of contexts more closely resembling those found in the 'real world'. This chapter outlines the philosophy of approach taken by researchers within NEnet in their recent efforts to pursue these aims.

Theories, Methods, Collaborations

While the demand grows for collaborations between neuroscience and education that embrace expertise and concepts from both perspectives, such collaborations are not straightforward. One fundamental issue is the significant philosophical divide between perspectives. Educational research, with its roots in social science, places strong emphasis upon the importance of human development, social context and the interpretation of

Educational Neuroscience, First Edition. Edited by Kathryn E. Patten and Stephen R. Campbell.
Chapters © 2011 The Authors. Book compilation © 2011 Educational Philosophy and Theory/Blackwell Publishing Ltd.
Published 2011 by Blackwell Publishing Ltd.

meaning. Neuroscience, on the other hand, is more concerned with controlled experimental testing of hypotheses and the identification of cause-effect mechanisms that can be generally applied. Concepts and language also differ widely, even with respect to the meaning of fundamental terms such as 'learning'. In cognitive neuroscience, learning is often synonymous with general memory abilities at the level of the individual. These include declarative memory, such as our ability to explicitly recall facts, but also non-declarative forms of memory such as the acquisition of skills and habits, conditioned emotional responses and even habituation to a repeated stimulus (Squire, 2004). Educators, on the other hand, more often describe learning in terms of social construction, through authentic exploration, engaging activities, interactive group work and student ownership of the learning process, emphasizing the importance of context. Additionally, educators may consider learning as closely bound to issues of meaning, the will to learn, values and the distributed nature of these and other aspects of learning beyond the level of the individual (TLRP, 2006; 2007).

These differences represent a major challenge for researchers at the interface between neuroscience and education and there may be no single solution to bridging them. Instead, one may expect some diversity in the approaches taken by emerging centres of research activity. Discussion within NEnet, which has been heavily influenced by the ESRC-TLRP seminar series, has given rise to a 'levels of action' model to help examine the potentially complex interrelationship between the different learning philosophies that meet in this emerging new field (Howard-Jones, 2008b; 2010, pp. 79–97). This model also suggests how the different methodologies associated with educational research and neuroscience can be usefully interrelated in neuroeducational research. The model builds on the brain->mind->behaviour model of cognitive neuroscience (Morton & Frith, 1995) and extends it to place greater emphasis on social processes. In Figure 1, the representation of two individuals interacting helps remind us of the complexity that can arise when processes more often studied at an individual level operate within a social environment.

The two individuals in Figure 1 may be two learners or, perhaps, a teacher and learner. In this diagram, the space between the individuals is filled by a sea of symbols representing human communication in all its forms. The lines separating brain, mind, behaviour and this sea of symbols are shown as dotted, to emphasise the somewhat indistinct nature of the boundaries between them and the difficulty in considering these as separable concepts. This levels-of-action model helps maintain awareness of the usefulness and limitations of different perspectives on learning. For example, work within NEnet has included an fMRI study of creativity fostering strategies (Howard-Jones *et al.*, 2005). This imaging study, which included a focus on the biological correlates of creativity, was useful in revealing how those parts of the brain associated with creative effort in a story telling task were further activated when unrelated stimulus words had to be included. Results provided some helpful indication, at the biological level of action, of the likely effectiveness of such strategies in the longer term. For example, had no such increased activity been observed, this might suggest that such strategies, although known to bring about outcomes that are judged as more creative, may do so without additional rehearsal of the processes regarded as creative. However, that cannot be the end of the story for educators. Taken in isolation, the study provides a poor impression of the issues involved in effectively implementing these strategies in the classroom. When and how

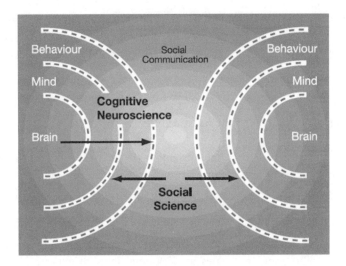

Figure 1: To interrelate the most valuable insights from cognitive neuroscience and the social science perspectives of education (represented by arrows), the brain->mind->behaviour model may need to be socially extended. Even two individuals interacting, as represented here, is suggestive of the complexity that can arise when behaviour becomes socially mediated. Such complexity remains chiefly the realm of social scientists, who often interpret the meaning of such communication in order to understand the underlying behaviour. Cognitive neuroscience has established its importance in understanding behaviour at an individual level but is only just beginning to contemplate the types of complex social domains studied by educational researchers

should they be used? To understand these issues, real world contexts must be meaning-fully interpreted. However, the meanings ascribed to actions of students and teachers in the classroom, including their use of language, are multiple, ambivalent and transitory. Although the production and perception of language have been fruitful areas for laboratory-based scientific research, the interpretation of meaning within everyday con-texts is essentially a problematic area for experimental scientific paradigms. Interpreta-tions of meaning that cannot be judged by the methods of natural science may be considered beyond its jurisdiction (Medawar, 1985). The recent flourishing of journals focusing on social cognitive neuroscience may demonstrate accelerating progress in this area, but interpretation of social complexity remains chiefly the realm of social scientists. Rather than natural science then, it is social science, with its own concepts of reliability and validity, that appears most accomplished in interpreting meaning at the social level of action, in order to understand the fuller significance of human communication (Alexander, 2006). Such considerations, in the context of our research on creativity, prompted an action research project in which an interdisciplinary team and a group of trainee teachers co-constructed concepts about the fostering of creativity that were both scientifically valid and educationally relevant. This subsequent study highlighted the importance of teachers' broader awareness of cognition and brain function in imple-menting such strategies (Howard-Jones *et al.*, 2008). Here, experiential accounts and meaning-based interpretations of discourse were useful in understanding the factors influencing pupils' creative progress, and how these might relate to concepts of brain and mind. Our qualitative work drew on educators' personal and classroom experience, together with findings from the fMRI study and further research from cognitive psychology and neuroscience that made these findings scientifically meaningful. It provided useful insights about pedagogical practice and how decisions to apply particular strategies should take into account the learner(s), their progress and the specific educational context (Howard-Jones, 2008a).

It would appear that neither natural nor social science, in isolation, presently offer sufficient epistemological traction to travel across all the levels of description shown in Figure 1. In this diagram, two of the most frequently travelled pathways of investigation associated with different perspectives are depicted by arrows. Cognitive neuroscience is shown to extend from brain to behaviour but little further, reflecting its present difficulties in penetrating complex, meaning-based social interaction. However the role of cognitive neuroscience is essential, as in our fMRI study of creativity, for supporting careful consideration of individual brain-mind relationships with biological and psychological evidence, and improving understanding of teaching and learning strategies at these levels of action. When it comes to a fuller understanding of how such interventions are applied in specific contexts, issues at the social level of action, such as individual differences in teachers' interactions with children, require exploration from social science perspectives more familiar to educational researchers. It is an interesting exercise to imagine how pathways associated with other perspectives might be depicted on this diagram. For example, phenomenological perspectives, that emphasise the role of mental reflection in understanding our own and others' behaviour, might be shown as an arrow from mind to the sea of symbols. (However, note how this diagram gives greater weight to 'outsider' perspectives and this can limit its helpfulness in representing perspectives that draw on experiential evidence and illuminate important issues of human agency- see below.)

Challenges, Results and Implications

Philosophy investigates the 'bounds of sense: that is, the limits of what can coherently be thought and said' (Bennett & Hacker, 2003, p. 399). Those who attempt to work at the interface of neuroscience and education will find themselves straddling at least two, very different, philosophies about learning, each expounding a very different set of concepts. Here, researchers are faced with the challenge of using language and developing new concepts that reside clearly within the bounds of sense as interpreted by both of these communities. Examples of common 'sense' transgressions in this new area are prevalent (Geake, 2008), but many appear linked to two extreme constructions of the mind-brain relationship.

In the extreme dualist camp, some educational issues are in danger of becoming entirely 'medicalised'. When educational issues become associated with biological issues, they can sometimes be characterised as entirely biologically determined and so removed from educators' domain of influence. One example is management of the increasing numbers of pupils considered to have challenging developmental disorders such as ADHD. Here, the increasing use of psychoactive drugs and images of differences in brain activity can lead to an increased sense of biological determinism (Degrandpre, 1999), and thus a diminished sense that outcomes are amenable to educational intervention. Another example might be the tendency for debates around dyslexia to be unhelpfully characterised as two options in conflict, a type of 'all-or-none' theorising that either dyslexia is a mental construction or derives entirely from a biologically determined cause (Nicolson, 2005). In the extreme monist camp, of course, such debates are meaningless, since mind and brain are conflated, and psychological and biological concepts are not distinguished. This monist departure from 'sense' is often found in popular language (e.g. 'my brain is confused') or in 'brain-based' educational programmes where synaptic connections become confused with psychological ones (see Wolfe, 1998) discussed by (Davis, 2004; Howard-Jones, 2008b).

At the root of such misunderstanding is the fact that interrelating mind and brain is not straightforward. Indeed, a whole field of scientific research has been founded on efforts to achieve such understanding. In the field of cognitive neuroscience, researchers believe that mind and brain must be explained together (Blakemore & Frith, 2000). The notion of mind is regarded as a theoretical but essential concept in exploring the emerging relationship between our brain and our behaviour, including our learning. Seen in this way, the study of cognition appears as a vital bridge in linking our knowledge of the brain to observations of behaviours, including those that involve learning. For this reason, it has been pointed out that without sufficient attendance to suitable cognitive psychological models, neuroscience will have little to offer education (Bruer, 1997).

The levels-of-action model incorporates the brain-mind-behaviour of cognitive neuroscience and thus helps avoid the dangers of dualist and monist notions. However, it remains flawed in another important sense. Educators are encouraged to develop autonomous learners, personally motivated and able to learn in response to their own free will. Indeed, effective teaching and learning is considered by many to depend upon the promotion of learners' independence and autonomy (e.g. see TLRP, 2007, p. 9). Some researchers within neuroscience, on the other hand, are presently unsure how, and even whether, free will comes into existence. Studies of apparent mental causation suggest that unperceived causes of action fail to influence our experience of will, suggesting that conscious will is an illusion: just the mind's way of estimating its own *apparent* authorship by drawing causal inferences about relationships between thoughts and actions (Wegner, 2003). This can be considered as another type of biological privilege likely to cause conflict for those working at the interface between neuroscience and education (Giesinger, 2006). However, such debates are not confined to education, since denying the existence of free will brings neuroscience into conflict with the entire legal system (Burns & Bechara, 2007). These arguments are bound up with those about consciousness and are unlikely to be resolved in the near future (Tancredi, 2007), allowing educators and other professionals to continue sharing, to a greater or lesser extent, a firmly held assumption that free will is a major causal factor in much behaviour.

At present, and perhaps reflecting again its close relationship to consciousness, it is not easy to represent human agency in Figure 1. However, given the growing emphasis on learning autonomy in education, researchers working at the interface between neuroscience and education must remain mindful that learning is best represented as a dynamic scenario in which change can include transformation of learners' (and teachers') own biological processes, perceptions and interpretations of meaning. The role of free will and reflexive self-determination may be powerful and essential contributions to learning that require careful consideration at all the levels represented here (biological, cognitive, behavioural and social). Returning to the particular example of our trainee drama teachers trying to foster creativity, we found it useful to carry out experiential arts-based workshops with professional actors prior to beginning the action research cycle, to gain insights related to free-will and ethical issues regarding control, through reflection with the artistic team (Howard-Jones & Border Crossings, 2005). This also generated video footage that supported subsequent discussion of these and other concepts with the trainee teachers.

Work within NEnet has successfully produced both scientific knowledge and practical educational understanding that can be implemented in the classroom. However, it has also drawn attention to a number of serious challenges for workers at the interface of neuroscience and education. Here, researchers must traverse the boundaries of diverse traditions of knowledge making and establish coherent interdisciplinary dialogue, maintaining 'sense' as it is commonly determined and understood by these very different traditions.

The emergence of a field of enquiry at the interface between neuroscience and education is generating a new dialogue around some very fundamental issues. It has not been possible to explore these issues in full within this short chapter. Instead, it is hoped that some sense of the challenge for workers in this area has been outlined. In addition, the chapter has attempted to convey the need for these challenges to be met with a creative, positive, critical and educationally grounded re-examination of how the diverse perspectives on learning offered by the natural and social sciences may be usefully interrelated.

Note

1. Our network prefers the term 'neuroeducational' rather than educational neuroscience, as we believe this better reflects a field with education at its core, uniquely characterised by its own methods and techniques, and which constructs knowledge based on experiential, social and biological evidence.

References

Alexander, H. A. (2006) A View from Somewhere: Explaining the paradigms of educational research, *Journal of Philosophy of Education*, 40:2, pp. 205–21.

Bennett, M. R. & Hacker, P. M. S. (2003) *Philosophical Foundations of Neuroscience* (Hoboken, NJ, Blackwell Publishing).

Blakemore, S. J. & Frith, U. (2000) *The Implications of Recent Developments in Neuroscience for Reseach on Teaching and Learning*. Report compiled for the ESRC Teaching and Learning Research Programme (Exeter, TLRP).

Bruer, J. (1997) Education and the Brain: A bridge too far, *Educational Researcher*, 26:8, pp. 4–16.

Burns, K. & Bechara, A. (2007) Decision Making and Free Will: A neuroscience perspective, *Behavioral Sciences and the Law*, 25:2, pp. 263–80.

Davis, A. J. (2004) The Credentials of Brain-Based Learning, *Journal of Philosophy of Education*, 38:1, pp. 21–36.

Degrandpre, R. (1999) Just Cause? *The Sciences*, 39:2, pp. 14–18.

Geake, J. G. (2004) Cognitive Neuroscience and Education: Two-way traffic or one-way street? *Westminster Studies in Education*, 27:1, pp. 87–98.

Geake, J. G. (2008) Neuromythologies in Education, *Educational Research*, 50:2, pp. 123–133.

Giesinger, J. (2006) Educating Brains? Free-will, brain research and pedagogy, *Zeitschrift fur erziehungswissenschaft*, 9:1, pp. 97–109.

Howard-Jones, P. A. (2007) *Neuroscience and Education: Issues and opportunities*, TLRP Commentary (London, Teaching and Learning Research Programme).

Howard-Jones, P. A. (2008a) *Fostering Creative Thinking: Co-constructed insights from neuroscience and education* (Bristol, Escalate).

Howard-Jones, P. A. (2008b) Philosophical Challenges for Researchers at the Interface between Neuroscience and Education, *Journal of Philosophy of Education*, 42:3–4, pp. 361–380.

Howard-Jones, P. (2010) *Introducing Neuroeducational Research* (Abingdon, Routledge).

Howard-Jones, P. A., Blakemore, S. J., Samuel, E., Summers, I. R. & Claxton, G. (2005) Semantic Divergence and Creative Story Generation: An FMRI Investigation, *Cognitive Brain Research*, 25, pp. 240–50.

Howard-Jones, P.A. & Border Crossings (2005) Creativity, Performance and the Brain – a Sci-Art Lab Workshop: 31 August-1 September.

Howard-Jones, P.A., Winfield, M. & Crimmins, G. (2008) Co-Constructing an Understanding of Creativity in the Fostering of Drama Education that Draws on Neuropsychological Concepts, *Educational Research*, 50:2, pp. 187–201.

Medawar, P. (1985) *The Limits of Science* (Oxford, Oxford University Press).

Morton, J. & Frith, U. (1995) Causal Modelling: A Structural Approach to Developmental Psychopathology, in: D. Cicchetti & D. Cohen (eds), *Manual of Developmental Psychopathology* (New York, Wiley), pp. 357–90.

Nicolson, R. (2005) Dyslexia: Beyond the Myth, *The Psychologist*, 18:11, pp. 658–59.

OECD (2007) *Understanding the Brain: Birth of a New Learning Science* (Paris, OECD).

Pickering, S. J. & Howard-Jones, P. (2007) Educators' Views on the Role of Neuroscience in Education: Findings from a study of UK and international perspectives, *Mind, Brain and Education*, 1:3, pp. 109–13.

Squire, L. R. (2004) Memory Systems of the Brain: A brief history and current perspective, *Neurobiology of Learning and Memory*, 82, pp. 171–77.

Tancredi, L.R. (2007) The Neuroscience Of 'Free Will', *Behavioral Sciences and the Law*, 25:2, pp. 295–2007.

TLRP (2006) Improving Teaching and Learning in Schools: A commentary by the Teaching and Learning Research Programme (London, TLRP).

TLRP (2007) Principles into Practice: A Teacher's Guide to Research Evidence on Teaching and Learning (London, TLRP).

Wegner, D. M. (2003) The Mind's Best Trick: How we experience conscious will, *Trends in Cognitive Sciences*, 7:2, pp. 65–69.

Wolfe, P. (1998) Revisiting Effective Teaching, *Educational Leadership*, 56:3, pp. 61–64.

5

What Can Neuroscience Bring to Education?

MICHEL FERRARI

Looking back over the mainstream relationship between psychology and education, we seem to see a clear progression. At the turn of the 20th century we had educational psychology emerging as a broad discipline to apply psychological findings to educational practice—although already we find a tension between the pragmatic approach of Dewey and the behaviorist 'drill and skill' approach of Thorndike. By the 1960s we see the first efforts to apply cognitive psychology to education, both in information-processing models and in constructivist efforts like those of Piagetians and neo-Piagetians. By the early 1990s we find efforts to consider the importance of contextual variables like family and culture, or learning environment. By the late 90s, we find models like those of Fischer (Fischer & Bidell, 2006) that try to integrate all of these approaches into a comprehensive model of knowledge development. As the 21st century begins, we have a new development, educational neuroscience.

Educational neuroscience promises to incorporate emerging insights from neuroscience into education and is an exciting renovation of cognitive science, if integrated with sensitivity. Although educational neuroscience aims to explain learning and development generally, it seems particularly important to explaining atypical performances of students with special needs (e.g. ADHD, autism, dyslexia, or math disabilities). However, for a full understanding of these disabilities, educational neuroscience must consider how they manifest themselves in learning and education that occur within the broad context of people's lives, both inside and outside of school. My own work and that of my students suggests that although educational neuroscience can advance our understanding of how knowledge is embodied, it needs to do so in ways that promote personal learning and development. For example, Carrie Richardson (2002) and I looked at how students with Attention Deficit Hyperactivity Disorder (ADHD) managed the intake of Ritalin to control their attentional difficulties. In a series of six case studies, we found that adolescents made personally and contextually sensitive decisions about when to take Ritalin, depending on the kind of experiences that they expected to encounter. For routine school tasks they would take the drug, but for creative tasks like composing music they would be careful to let its effect wear off before engaging in that activity. Studying the neuroscience underlying attentional difficulties certainly allows us to better to understand what areas of the brain (and what cognitive subskills) are involved in cognitive performances, and where points of

Educational Neuroscience, First Edition. Edited by Kathryn E. Patten and Stephen R. Campbell.
Chapters © 2011 The Authors. Book compilation © 2011 Educational Philosophy and Theory/Blackwell Publishing Ltd.
Published 2011 by Blackwell Publishing Ltd.

breakdown can contribute to ADHD, but what our study shows is that we must also understand the attentional demands of particular personally meaningful educational contexts (including both classroom and informal learning contexts). Likewise, Zopito Marini and his colleagues (Marini *et al.*, 2010) have also shown that learners who show similar kinds of bullying and aggression can require very different kinds of scaffolding or training when those behaviours occur for different neurobiological reasons that lead some people to be callous and others to be uninhibited; understanding these differences can allow educators to react more sensitively to students who engage in bullying, and to be much better prepared to help them manage their aggression.

All this seems a very promising use for educational neuroscience by educators, and many seem very excited about the possibilities of what neuroscience can bring to education. However, as Geake and Cooper (2003a, b) point out, there are also dangers in efforts to apply neuroscience to education. One danger is when neuroscience adds nothing new to existing understanding of the issues, but lowers one's guard about the value of particular kinds of educational practice—simply riding a bandwagon and the wave of popularity of neuroscience explanations generally. Indeed, a recent study showed that explanations with neuroscience included seemed better, even when neuroscience added nothing in support of the arguments presented (Weisberg *et al.*, 2008). As a real-life case in point, although there are clear neurocognitive differences between people diagnosed with autism and the rest of the population, what these differences mean for autistic individuals as they live their lives outside the lab is not immediately apparent. Indeed, lab study results from neuroscience and cognitive science can often lead to false generalizations from lab situations to people's lives. For example it is sometimes reported that people with autism have no understanding of other minds—that they are, in a sense, 'mind-blind' (Baron-Cohen, 1995). Ljiljana Vuletic and I conducted an in-depth case study of Teodor Mihail, an adolescent with Asperger's Syndrome (Vuletic, Ferrari & Mihail, 2005)—which many place on the Autism spectrum—and found that he showed great sensitivity and insight into other people's condition when those people were significant to him. For instance, he was concerned that his grandmother might become too tired if they took too extensive a bus tour, despite his own fascination with the Toronto bus system; and he had no trouble understanding that people in Romania (where he had visited) could not know about Toronto buses because they had never been to Toronto.

Another danger of uncritically applying neuroscience to educational practice is reductionism. We must guard against claiming that the root cause of learning difficulties is a mechanical failure or abnormality that operates at the genetic or neural level. It is important to acknowledge that the brain changes itself in response to environmental influences and based on personal effort and choices (Doidge, 2007). This is a point Bandura (2006) makes in his critique of studies in *cognitive* neuroscience that seem to undermine human agency. Following Bandura, it is important that *educational* neuroscience be careful to promote frameworks in which agency is possible and valued.

We can see the importance of agency in the examples, mentioned earlier, of students determining how to manage the use of Ritalin. Likewise, in our case study of Asperger's Syndrome, Teodor later asked Ljiljana what most adolescents would put on a website, so he could be sure to do the same. Not only atypical adolescents such as these, but the people who care about them—friends and parents—all work to position these youth in light of, and sometimes as resisting, master narratives about disabilities that are culturally very powerful and as formative for their lives as any neuroscientific evidence concerning what is or is not possible for them to do (Hammack, 2008; Harré, 2008; Harré & van Langenhove, 1999). This point is particularly clear when we consider people with Asperger's who have left school and function as adults in society—sometimes (but not always) as successfully as anyone else, as Ljiljana's interviews with adults with Asperger's clearly show (Vuletic, 2010).

Thus, we endorse the Dalai Lama's (2006) view that although mind and self are empty of inherent existence, that does not mean that there is no lived experience of mind or self; rather it means that to fully understand them, one needs to understand the causes and conditions of their existence. While neuroscientific evidence points to an important role for 'upward' neural causes for various kinds of experience, the kind of mind or person we become is also 'downwardly' caused by our education and our choices (Sperry, 1993). Of course, not everything about our mind or person is a product of our education, even very broadly defined. The physical environment has as important an effect on the brain as the brain has on our capacity for learning. As Martha Farah (2010) has shown in a series of studies, low socioeconomic status (SES) has dramatic effects on the brain due to associated physical effects of malnutrition, stress, as well as the psychological effects of lack of cognitive stimulation; hence, we need to be alert to powerful impact of poverty on education through its direct effect on neurocognitive development, and work to mitigate that impact before low SES produces lasting effects on children's brains and minds that education can only partially remediate. In that light, educational neuroscience studies such as those of Farah (2010) can provide evidence in support of something as basic as the importance of providing nutritious lunches at school—which can be critical in assuring healthy brain development. Educational neuroscience can also help design educational programs, for example, based on observed differences in learning success and brain functioning of those more or less skilled in math.

How, then, to proceed to make the most efficient use of the new information gained by studies in educational neuroscience and their meaning for people's lives? Ironically, history gives us good models that have not been maximally exploited to date. In particular, the original writings of both Piaget (1967, 1983 [1970]) and Vygotsky (1997) [1934] show that they had already incorporated the neuroscience of their day into overarching research programs that considered their implications for education in ways that might be adopted in outline today. For instance, Piaget proposed a reciprocal assimilation of the findings from biology and cognitive science (as well as other allied disciplines). While appreciating the distinction between causal explanations in neurobiology and implicative explanations in psychology, he granted a certain iso-morphism between the kinds of structures and their relation to learning activities (Ferrari, 2009).

Furthermore, although Vygotsky has little to say about isomorphic knowledge struc-tures and how they relate to specific brain activity, he had much more to say about the relations between culturally developed bodies of knowledge and the individuals who must learn them. In particular, he was deeply concerned with how people with physical deficits (e.g. those born with handicaps like blindness or deafness, or acquired injuries such as brain damage) must learn through alternative pathways—a point echoed by Fischer and Bidell (2006). I wholeheartedly endorse this view, but would like to add that we must also include technology and the extended environment when considering the scope and limitations of educational neuroscience. Consider, for instance, Vaughan, Rogers, Singhl and Swalehe's (2000) report on how collective self-efficacy surrounding family planning and sexual responsibility can be promoted through educational entertainment. In particular, a specially designed radio show significantly increased HIV/AIDS prevention in the part of Tanzania in which it aired.

In sum, we need an educational neuroscience that follows the medical model at least in this way: that pure research informs practice, especially for rare cases that deviate from the norm. But unlike medicine, which aims at promoting health, education promotes values that reflect the kind of citizen and ultimately the kind of society we aspire to create. Although educational neuroscience necessarily involves evidence from cognitive neuroscience about the brain, it also concerns people and how they choose to live their lives, as shaped by the cultural influences they are exposed to. This point is wonderfully articulated in a debate between the neuroscientist Jean-Pierre Changeux and the philosopher Paul Ricœur (Changeux & Ricœur, 2000) about whether neuroscience can enter into a 'third discourse' that unites biological and personal experience. However, this debate did not specifically address the issue of education.

Still, the issue of creating a common discourse seems relevant to this debate about how to integrate education and neuroscience, at least to the extent that one agrees with Egan (1997), who proposes that there are at least three aims of public education: (1) job preparation; (2) truth seeking and (3) personal flourishing. All three of these may not only conflict with each other, they may also call for different relations between education and neuroscience, depending on what we choose to include within them. In other words, neuroscience will inform each of these three aims of education, but does not set any aims itself; educators or policy makers must look to neurobiology to answer the specific questions that concern them personally, questions such as 'What is the biological basis of personal flourishing?' or 'What biological support is needed to allow particular people to learn in ways that will get them a good job?'. Existing research programs in educational neuroscience that aim to help promote literacy and numeracy might be an important part of any answer to such questions, but different questions suggest different aspects of our biology may need to be explored to answer them. What are the biological foundations of authentic and deep understanding? Of an appreciation of art and beauty? Or of compassion for those in need at home and around the world? All these concerns reflect different values that matter to particular communities and neuroscience could inform us about all of them.

Thus, again, as cases in point, students with ADHD or Asperger's certainly have trouble with certain aspects of the contemporary educational context—like sitting still for long periods of time performing tasks deemed important in our school curriculum. Such difficulties can be traced to features of their underlying neurochemistry (Dalley *et al.*, 2008), but those features and those difficulties are neither the last word about them as people, nor even about what they are able to learn under other socio-cultural conditions, nor what other kinds of learning they and others might consider a valued contribution to society.

True, neuroscience cannot study everything, but then, neither can all learning of interest fit into a K-12 curriculum with only limited time and resources. Thus, educational neuroscience must itself become part of a broader debate about the aims of education and how to help students flourish, understand deeply, and become socially productive members of society. Perhaps learning basic skills like reading and mathematics is not enough, and we also need to encourage students to gain self-insight (or personal wisdom) through practices such as mindfulness meditation (Rosch, 2008). If so, and recent efforts at promoting contemplative education alongside the basic curriculum become successful, then contemplative neuroscience will suddenly become as central to educational neuroscience as are efforts to understand the neurocognitive basis of dyslexia and dyscalculia.

In other words, educational neuroscience can help fulfil the mandate of public education, but only as a tool that is part of a broader conversation in Canada, and in other countries around the world, about what schools should strive to achieve for the millions of students who attend them.

References

Bandura, A. (2006) Toward a Psychology of Human Agency, *Perspectives on Psychological Science*, 1:2, pp. 164–180.

Baron-Cohen, S. (1995) *Mindblindness: An essay on autism and theory of mind* (Boston, MA, MIT Press/Bradford Books).

Changeux, J. P. & Ricœur, P. (2000). *What Makes Us Think?* (Princeton, NJ, Princeton University Press).

Dalai Lama (2006) *How to See Yourself as You Really Are* (New York, Atria Books).

Dalley, J. W., Mar, A. C., Economidou, D. & Robbins, T. W. (2008) Neurobehavioral mechanisms of impulsivity: Fronto-striatal systems and functional neurochemistry, *Pharmacology, Biochemistry and Behavior*, 90, pp. 250–26.

Doidge, N. (2007) *The Brain That Changes Itself: Stories of personal triumph from the frontiers of brain science* (New York, Viking).

Egan, K. (1997) *The Educated Mind: How cognitive tools shape our understanding* (Chicago, IL, University of Chicago Press).

Farah, M. J. (2010) Mind, Brain and Education in Socioeconomic Context, in: M. Ferrari & L. Vuletic (eds), *Developmental Interplay of Mind, Brain, and Education: Essays in honor of Robbie Case* (Dordrecht, Springer).

Ferrari, M. (2009) Piaget's Enduring Contribution to a Science of Consciousness, in: U. Mueller, J. Carpendale & L. Smith (eds), *Cambridge Companion to Piaget* (Cambridge, Cambridge University Press).

Fischer, K. W. & Bidell, T. R. (2006) Dynamic Development of Action, Thought, and Emotion, in: R. M. Lerner (ed.), *Handbook of Child Psychology. Vol. 1. Theoretical models of human development*, 6th edn. (New York, Wiley).

Geake, J. & Cooper, P. W. (2003a) Cognitive Neuroscience: Implications for education? *Westminster Studies in Education*, 26, pp. 7–20.

Geake, J. & Cooper, P. W. (2003b) The Educated Brain: The relevance of cognitive neuroscience to educational theory and practice, *Westminster Review of Educational Studies*, 26, pp. 7–20.

Hammack, P. L. (2008) Narrative and the Cultural Psychology of Identity, *Personality and Social Psychology Review*, 12, pp. 222–247.

Harré, R. (2008) Positioning Theory, *Self-Care & Dependent-Care Nursing*, 16, pp. 28–32.

Harré, R. & van Langenhove, L. (1999) *Positioning Theory* (Oxford, Blackwell).

Marini, Z. A., Dane, A. V. & Kennedy, R. E. (2010) Multiple Pathways to Bullying: Educational Implications of Individual Differences in Temperament and Brain Function, in: M. Ferrari & L. Vuletic (eds), *Developmental Interplay of Mind, Brain, and Education: Essays in honor of Robbie Case* (Dordrecht, Springer).

Piaget, J. (1967) *Biologie et connaissance* [Biology and Knowledge] (Paris, Gallimard).

Piaget, J. (1983) [1970] Piaget's Theory, in: P. H. Mussen (series ed.), *Handbook of Child Psychology: Vol. 1.* W. Kessen (ed.) *History, Theory, and Methods*, 4th edn. (New York, Wiley), pp. 103–128.

Richardson, C. A. (2002) *A Look at Adolescent Attention Deficit/Hyperactivity Disorder Form the Inside: How medication is perceived to affect one's sense of self.* Unpublished Master's thesis, University of Toronto.

Rosch, E. (2008) Beginner's Mind: Paths to the wisdom that is not learned, in: M. Ferrari & G. Potworowski (eds), *Teaching for Wisdom* (Dordrecht, Springer), pp. 135–162.

Sperry, R. W. (1993) The Impact and Promise of the Cognitive Revolution, *American Psychologist*, 48, pp. 878–885.

Vaughan, P. W., Rogers, E. M., Singhl, A. & Swalehe, R. M. (2000) Entertainment-education and HIV/AIDS Prevention: A filed experiment in Tanzania, *Journal of Health Communication*, 5 (supplement), pp. 81–100.

Vuletic, L., Ferrari, M. & Mihail, T. (2005) *Transfer Boy: Pespectives on Asperger syndrome* (London, Jessica Kingsley Press).

Vuletic, L. (2010) *Adults Living with Asperger's Syndrome.* PhD Thesis, University of Toronto.

Vygotsky, L. S. (1997) [1934] Psychology and the Theory of Localization of Mental Functions, in: R. W. Rieber & J. Wollock (eds), *The Collected Works of L. S. Vygotsky. Vol. 3, Problems of the Theory and History of Psychology* (New York, Plenum Press), pp. 139–144.

Weisberg, D. S., Keil, F. C., Goodstein, J., Rawson, R. & Gray, J. R. (2008) The Seductive Allure of Neuroscience Explanations, *Journal of Cognitive Neuroscience*, 20, pp. 470–477.

6

Connecting Education and Cognitive Neuroscience: Where will the journey take us?

Daniel Ansari, Donna Coch & Bert De Smedt

Introduction

There has been tremendous growth in the scientific study of the human brain over the last 15 years, and a concomitant excitement surrounding new findings about how the brain works. The burgeoning availability of non-invasive tools and techniques used to measure brain function during cognitive tasks led to the creation of the field of Cognitive Neuroscience in the early 1990s, and the continuous development of such tools has supported the remarkable growth of this field since then. Broadly speaking, the aim of Cognitive Neuroscience is to elucidate how the brain enables the mind (Gazzaniga, 2002). In other words, the goal of Cognitive Neuroscience is to constrain cognitive, psychological theories with neuroscientific data, thereby shaping such theories to be more biologically plausible. Throughout the 'Decade of the Brain' in the 1990s and into the 21st century, cognitive neuroscience research has been widely popularized, with colorful brain images filling the 'Science and Nature' sections of daily newspapers frequently.

Recently, research in cognitive neuroscience has attracted the attention of educationalists. Naturally, people interested in learning and education might want to know how results from relevant cognitive neuroscience research could be applied in the classroom. Given that the brain is the 'organ of learning' it seems logical that knowledge about how the brain works should be able to inform education. Indeed, there is a growing body of cognitive neuroscience research in areas that are of potential key interest to education, such as research on the neural correlates of reading (e.g. Pugh *et al.*, 1996; Turkeltaub *et al.*, 2003) and mathematics and number processing (e.g. Dehaene *et al.*, 2003; Dehaene *et al.*, 1999). Such research has generated great hopes amongst some for a revolution in education in which results from the neuroscience laboratory positively transform the classroom. In fact, a number of journal publications have reviewed and discussed evidence from cognitive neuroscience that might be relevant to education (Ansari & Coch, 2006; Blakemore & Frith, 2005; Goswami, 2004, 2006; Posner & Rothbart, 2005; Stern, 2005). Further evidence for the growth in excitement surrounding the potential connection between neuroscience and education is the creation of a new international society for Mind, Brain, and Education (http://www.imbes.org/) accompanied by the launch of a peer-reviewed journal of the same name. Of course,

Educational Neuroscience, First Edition. Edited by Kathryn E. Patten and Stephen R. Campbell.
Chapters © 2011 The Authors. Book compilation © 2011 Educational Philosophy and Theory/Blackwell Publishing Ltd.
Published 2011 by Blackwell Publishing Ltd.

the present book is further testament to growth in the emerging field of Mind, Brain, and Education.

These are certainly exciting times, in which the potential for real and meaningful connections between education and cognitive neuroscience is strong and widely supported (although cf. Bruer, 1997). However, in the context of this wave of excitement surrounding such potential connections, a number of questions have been left largely unconsidered in a systematic fashion: exactly how will cognitive neuroscience inform education, and how will education inform cognitive neuroscience? At what levels will such connections be most fruitful, in terms of generating useable knowledge? What practical changes need to be undertaken in order to support such sustainable connections? What will be the roles of cognitive neuroscientists, educators, funding agencies, and policy makers in this endeavor? How will existing philosophical boundaries between so-called 'applied' (i.e. in the real world of the classroom) and 'basic' (i.e. in the controlled world of the laboratory) research be overcome?

We discuss some of these critical questions in the present chapter. We feel that careful consideration of these issues is necessary in order to facilitate and sustain connections between education and cognitive neuroscience. We contend that without concerted thinking about *how* to build bridges and maintain them, the very idea that cognitive neuroscience and education can interact to improve education will become just another educational fad, a footnote in the history of the movement towards research-based education. In considering these critical questions, we also discuss the potential future of the emerging field of Mind, Brain, and Education—where this journey will take us—and some important constraints that should guide us along the way.

How Might Cognitive Neuroscience Inform Education?

According to one perspective, the ideal connection between education and cognitive neuroscience would be as follows: cognitive neuroscientists would conduct experiments and then educators would directly apply the results of this research in their teaching; there would be a seamless flow from the laboratory to the classroom. Indeed, there are frequently calls for such direct links and the enterprise of Mind, Brain, and Education is considered by some to have failed if such links cannot be achieved. At a recent conference, one of us was asked what he would tell teachers to do on the basis of his research results. When he answered that he would like to first hear from teachers what *they* thought about the results and how *they* thought the findings might or might not be informative, there was visible disappointment on behalf of the questioner that a straightforward recipe derived from the research was not forthcoming. Here is an example of a philosophical divide that has plagued the history of education as a science; as Condliffe Lagemann notes: 'When educational scholarship was professionalized, it was viewed with contempt by noneducationalists; when it was discipline-based, it was shunned by students, who had wanted "recipes for practice"' (2000, p. 179).

We contend that such expectations for silver bullets, for research-based 'fixes' of educational problems, for easy-to-follow 'recipes for practice' based on cognitive neuroscience findings, are bound to be disappointed quickly; moreover, we argue that such expectations are unrealistic and threaten to erode efforts to forge useful connections between education and neuroscience. Indeed, there exists a growing industry of so-called 'brain-based learning' products that propose pedagogical approaches and introduce tools and teaching techniques that claim to be based on neuroscientific data. However, close inspection of these claims for a direct connection between particular 'brain-based' tools and teaching approaches reveals very loose and often factually incorrect links.

We do not believe that this sort of approach is the most fruitful for creating a sustainable science of Mind, Brain, and Education that mutually benefits education and cognitive neuroscience. Instead, we believe that the real potential lies in systematic interactions between cognitive neuroscientists and educators to arrive at common questions and a common language, rather than in the direct route from research to its application. The history of 'applied' research shows that the implementation of research results to solve problems is often very indirect and rarely straightforward. This is especially the case in Education, a field in which there has been much resistance to the potential influence of scientific, quantitative research (Lagemann, 2000). We expect that developing the field of Mind, Brain, and Education and the collaborations at the core of the field will require a journey much more complex than the direct route from the neuroscience laboratory to the classroom.

In light of this, the question that leads this section (how might cognitive neuroscience inform education?) is limiting inherently, as it contains the implicit assumption of a one-way link between education and neuroscience. Such unidirectionality is either implicitly or explicitly stated in discussions about education and neuroscience too often. We believe that instead of asking what neuroscience can do for education it should be asked how education and neuroscience might inform *each other*; that is, it is our explicit assumption that Mind, Brain, and Education should be framed in terms of interactions and based on mutually beneficial dialogue among participants with knowledge of child development, learning, and teaching. This will also ensure that no knowledge hierarchy is created in which educators are simply the recipients of information generated by neuroscientists. There is often a perception that scientists will tell educators what they should do; such a patronizing approach will be avoided if the types of collaborations we propose are realized.

As we have previously articulated (Ansari & Coch, 2006; Coch & Ansari, 2009), we argue that it is crucial for training in aspects of cognitive neuroscience to become a fundamental part of teacher education, while at the same time graduate students in cognitive neuroscience should be exposed to educational issues. We believe that such instructional components will help teachers to gain a fuller understanding of child development and the biological constraints placed on learning processes as well as research methodologies; similarly, cognitive neuroscientists investigating subjects that have potential educational relevance will be familiar with pedagogical issues surrounding their subject matter as well as with the related 'burning' questions being asked by educators and the constraints of the classroom learning environment. For example, educators might discuss with cognitive neuroscientists the different strategies that they have observed children using to solve a particular problem, or allow cognitive neuroscientists to observe children using various strategies in the classroom environment, thus providing an avenue to potentially bring some of the rich and deep descriptions of classroom learning into the neuroscience realm. This all will facilitate the generation of new interdisciplinary researchers fluent in the languages of education and cognitive neuroscience. In turn, this will result in collaborations from which new research questions will emerge that are closely aligned with both the traditional basic science interests of the cognitive neuroscientist and the applied issues encountered by teachers in their classrooms.

What Needs to Happen for Education and Neuroscience to Interact?

There are a number of practical issues that need to be addressed before interactions between educators and neuroscientists of the sort described above can become a reality. Here we focus on the teacher preparation issue mentioned above as an example. We believe that teacher education programs need to integrate courses on cognitive neuroscience into their curricula, or integrate cognitive neuroscience methods and findings into their current courses. Such courses should provide not only a basic introduction to structural and functional brain development as well as the brain mechanisms subserving core domains of cognitive functions such as the typical and atypical development of reading and mathematical skills, but also discuss wider topics of relevance to education such as the effects of culture on brain function. Of course, such courses should not be focused solely on results from brain imaging studies, but should also discuss evidence from behavioral research; by definition, Cognitive Neuroscience is an interdisciplinary science that draws on results from cognitive psychology, neuroscience, sociology, and anthropology to generate a better understanding of the brain bases of cognitive processes. In order to understand and better support human learning and development in their students, teachers need to know what science has discovered about learning and development at multiple levels of analysis, from multiple perspectives.

Teacher training should also introduce aspiring teachers to research methodologies, the strengths and limitations of behavioral methods and methods that measure brain activity, as well as the uses and misuses of scientific data in popular publications. Being able to critically evaluate scientific results and their portrayal in the popular media is crucial, especially because there already exists a great proliferation of so-called 'neuromyths' in publications aimed at teachers (for a review, see Goswami, 2004). As discussed above, there is a growing body of pedagogical tools and literature that claims to be 'brain-based'. Teachers who are able to critically evaluate the science to which they are being exposed will not only avoid heeding advice based on inaccurate data and pseudoscience, but also will force the producers of education-related literature on the brain to provide more sophisticated and accurate information. In other words, teachers need to become 'neuroscience literate'; and, by the same token, cognitive neuroscientists need to become 'education literate' in order for strong links to be forged between fields.

Thus, in order to forge such links, traditional academic boundaries need to be crossed and mutual respect developed, perhaps building on a common and shared foundational interest in child development and learning between developmental cognitive neuroscientists and educators. This will also require Departments of Education to lower their resistance to quantitative scientific, empirical research and, at the same time, Departments of Psychology and Neuroscience to embrace the importance of applied research, which is frequently considered inferior to the pursuit of knowledge characterized by basic research.

What Does the Future Hold?

This book on Education and Neuroscience comes at a critical time. There has been a steadily growing interest in the potential of a connection between Cognitive Neuroscience and Education. However, this interest may be reaching its peak and may soon subside, in part because the direct application of neuroscience findings to the classroom has not been particularly fruitful. It is therefore crucial to think about the ways in which the current enthusiasm and willingness of universities and funding agencies to engage with the creation of Mind, Brain, and Education as a new field can be sustained over the long-term. We contend that this can be achieved by moving beyond thinking about the direct application of neuroscience research results to classroom practice towards thinking about the constraints that need to be set in place in order to bring educators and neuroscientists together to collaborate and inform each others' thinking and practice. It will be important to communicate the potential and promise of such indirect links to policy makers, funding agencies, and universities in order to avoid their turning away from Mind, Brain, and Education when quick fixes are not forthcoming.

This view of Mind, Brain, and Education stands in stark contrast to much of the current 'brain-based' education movement. It is of concern that much energy will need to be expended in the future to curtail the growing emergence of so-called 'brain-based' programs and publications that proliferate myths throughout the educational community. In a similar vein, school boards and districts should be careful to choose and use only programs for which there is clear, peer-reviewed, empirical support regarding efficacy. In turn, it is important that scientists do not succumb to the temptation to collaborate with industry to create intervention tools that are only loosely based on their research results and have not undergone rigorous evaluation (particularly in classroom contexts) after initial publication simply in order to avoid potential commercial losses.

Finally, we believe that the future of Mind, Brain, and Education should be characterized by much broader thinking about how Neuroscience and Education might inform each other. What new research paradigms might be developed? How might non-invasive neuroimaging methods be used to measure the relative success of educational approaches? How can synergistic collaborations create a whole that is more than the sum of the parts? Moving beyond our understanding of the neurocognitive mechanisms subserving core cognitive domains, such as reading and mathematics, Mind, Brain, and Education also encompasses consideration of issues related to the general structure of learning environments, the timing of instruction, and the roles of stress, nutrition, sleep, and social context in learning (to name a few topics). Interactions between Education and Neuroscience may also help to evaluate the relative benefits of arts and science education and thereby change the way in which we view educational priorities. Where this journey will take us may be unpredictable *a priori*, but it is relatively more certain that we will not make strides without some constraints in place and a concerted effort to map out how we are going to get there.

References

Ansari, D. & Coch, D. (2006) Bridges Over Troubled Waters: Education and cognitive neuroscience, *Trends in Cognitive Sciences*, 10:4, pp. 146–151.

Blakemore, S. J. & Frith, U. (2005) *The Learning Brain: Lessons for education* (Oxford, Blackwell).

Bruer, J. T. (1997) Education and the Brain: A bridge too far, *Educational Researcher*, 26:8, pp. 4–16.

Coch, D. & Ansari, D. (2009) Thinking About Mechanisms is Crucial to Connecting Neuroscience and Education, *Cortex*, 45, pp. 546–7.

Dehaene, S., Piazza, M., Pinel, P. & Cohen, L. (2003) Three Parietal Circuits for Number Processing, *Cognitive Neuropsychology*, 20:3–6, pp. 487–506.

Dehaene, S., Spelke, E., Pinel, P., Stanescu, R. & Tsivkin, S. (1999) Sources of Mathematical Thinking: Behavioral and brain-imaging evidence, *Science*, 284, pp. 970–974.

Gazzaniga, M. S. (2002) *Cognitive Neuroscience*, 2nd edn. (New York, W. W. Norton & Company).

Goswami, U. (2004) Neuroscience and Education, *British Journal of Educational Psychology*, 74:Pt 1, pp. 1–14.

Goswami, U. (2006) Neuroscience and Education: From research to practice? *Nature Reviews Neuroscience*, 7, pp. 406–413.

Lagemann, E. C. (2000) *An Elusive Science: The troubling history of education research* (Chicago, IL, University of Chicago Press).

Pugh, K. R., Shaywitz, B. A., Shaywitz, S. E., Constable, R. T., Skudlarski, P., Fulbright, R. K., *et al.* (1996) Cerebral Organization of Component Processes in Reading, *Brain*, 119, pp. 1221–1238.

Posner, M. I. & Rothbart, M. K. (2005) Influencing Brain Networks: Implications for education, *Trends in Cognitive Sciences*, 9:3, pp. 99–103.

Stern, E. (2005) Pedagogy Meets Neuroscience, *Science*, 310, p. 745.

Turkeltaub, P. E., Gareau, L., Flowers, D. L., Zeffiro, T. A. & Eden, G. F. (2003) Development of Neural Mechanisms for Reading, *Nature Neuroscience*, 6:7, pp. 67–73.

7

Position Statement on Motivations, Methodologies, and Practical Implications of Educational Neuroscience Research: fMRI studies of the neural correlates of creative intelligence

JOHN GEAKE

My primary motivation in advancing educational neuroscience is to enable it, as an inter-disciplinary field, to become a mutually informative two-way street. Consequently, I offer this definition:

> *Educational neuroscience is cognitive neuroscience which investigates educationally inspired research questions.*

In other words, educational neuroscience is cognitive neuroscience that is relevant to, has implications for, might lead to applications in, educational practice and policy—pedagogy and curriculum—because the science addresses an educational problem or issue. Consequently, educational neuroscience, as a research endeavor, only makes sense if the genesis of its projects lies in educational issues, concerns, or problems. Without being rooted in education, neuroscientific data and interpretations are unlikely to be embraced by the education profession.

Therefore, educational neuroscience methodologies must necessarily incorporate an action research cycle, wherein the original educational issue inspires a set of cognitive neuroscientific research questions, which after investigation, the results have implications and applications for educational policy and/or practice. For the latter, the research cycle is not complete until the putative applications have been field tested in classrooms. Of course, the outcomes of this might lead to a revision of the articulation of the original educational issue such that a whole new raft of neuroscientific research questions arise, and another research cycle is initiated.

A key assumption is that a neuroscientific understanding of children's learning could enhance the professionalization of teachers. Presumably better-informed practice results in more efficient and enjoyable learning, which in turn should mean a better teaching experience for educators. To this end, some findings from educational neuroscience may support long-standing existing practice—so-called craft knowledge—thus providing a confidence buffer against what can seem to be an unrelenting tide of educational policy

Educational Neuroscience, First Edition. Edited by Kathryn E. Patten and Stephen R. Campbell.
Chapters © 2011 The Authors. Book compilation © 2011 Educational Philosophy and Theory/Blackwell Publishing Ltd.
Published 2011 by Blackwell Publishing Ltd.

change. However, it is also just as likely that some other findings from educational neuroscience will support calls for radical changes to educational practice (Geake & Cooper, 2003).

Our brains did not evolve to go to school. Yet we do all go to school as students, and some of us return as teachers. In recent years, neuroscientists have not been shy in publicizing new understandings of brain function in the considerable popular literature on the brain prominently displayed in all good booksellers. Curiously, whereas this literature features the brain basis of learning, memory, knowledge, even reading and mathematics, there is almost no mention of education, schools or classrooms. Equally curiously, in the vast mountains of educational policy, curriculum and outcomes documentation, there has been, until very recently, no mention of the human brain, the organ most central to the educational enterprise. It is as though education has been regarded as having little to do with how learning actually takes place in the brains of students. This is why I believe that educational neuroscience needs a discipline-specific methodology such as outlined above.

It could be noted, however, that the responses of the education profession to the emergence of educational neuroscience, especially in the UK, have been mixed. On the one hand, there are those aged education academics who, after a lifetime of not understanding and disparaging all science, see no need to change their ways now. On the other hand, there are the 'brain-based' enthusiasts who hope that the current fads of left–right thinking, brain gym, etc., will address the complexities and daily challenges of the mixed-ability classroom. A middle-way requires neuroscientific education for both groups so that the education profession can shape a professionally informative educational neuroscience research agenda of the future (Geake, 2005).

But, I suggest, the *prima facie* case that the education profession could benefit from embracing rather than ignoring cognitive neuroscience will fail unless educationists actively contribute to the future research agenda of brain research. That is, a cognitive neuroscience-education nexus should be a two-way street (Geake, 2004). Whereas cognitive neuroscience could inform education by providing additional evidence that confirms good practice, helps resolve educational dilemmas, or suggests new possibilities in pedagogy or curriculum design, education could inform cognitive neuroscience by providing a source of complementary behavioral data, especially on children, as well as posing new worthwhile lines of investigation. There are many eternal questions that teachers confront every day in the classroom that could be put onto the neuroscience research agenda, such as, why do some children learn more easily than others.

My collaborative educational neuroscience research has focused on trying to better understand how the brain enables creative thinking: how does the brain generate insights, crack jokes, compose wonderful tunes, solve difficult problems, think up interesting research questions, write poetry, and so on? A diverse research literature from education, cognitive psychology, artificial intelligence, archaeology, and cognitive neuroscience indicates that analogy-making is the key underpinning process of intelligence. Importantly, this is not analogy-making in the strict sense often used in intelligence testing, e.g. black is to white as night is to . . . ? Rather, creative thinking requires fluid analogizing, where fluid analogies are those without a strict or limited 'correct' answer. (For an extensive discussion of fluid analogizing, see Hofstadter, 1995; 2001.) A simple example is afforded by the question: What is the London of the USA? Many people will respond 'New York city', but politicians say 'Washington DC', film-makers could say 'LA', and geographers might reply 'London, Ohio'. The point is that none of the answers is 'wrong'; all are plausible. Most real-world categorizations are similarly fluid. In education, it has been argued that the effective employment of fluid analogies enables efficient epistemic categorization and thus assimilation of new knowledge (Geake & Dodson, 2005). Certainly, a characteristic of good teachers and scholars is their ability to readily create fluid analogies for explanation and clarification. Moreover, fluid analogizing has been conceptualized as a basic cognitive process underpinning those creative aspects of intelligence which are mediated through working memory (Geake & Hansen, 2005). Thus, using the methodological framework outlined above, the education issue behind our research is how to promote creative thinking in educational settings, while the consequent neuroscientific questions are about what role fluid analogizing plays in creative intelligence, and how the brain supports fluid analogical thinking.

The neuroscientific investigations have entailed using functional magnetic resonance imaging (fMRI) to map cortical locations where neural activity is higher when individuals are engaged in fluid analogizing problems (Geake & Hansen, 2005; 2006). These mappings reveal patterns of frontal-parietal cortical activations that are similar to those activation patterns found in previous studies into the neural correlates of high intelligence in general, and analogical reasoning in particular (Luo *et al.*, 2003; Wharton *et al.*, 2000). The conclusion is that fluid analogizing is a basic process for all cognitive endeavors.

More generally, intelligence requires neural systems which are distributed throughout rather than being restricted to specific locations in the brain. Consequently, simplistic models of intelligence such as multiple intelligences (Waterhouse, 2006) although popular in educational circles, should be quietly forgotten, together with the myriad of neuromythologies (neuro-nonsense?) such as only using 10%, left- and right-brain thinking, and visual, auditory and kinesthetic learning styles (Kratzig & Arbuthnott, 2006; Geake, 2008). All of these neuromythologies seem to have arisen from an ignorance of a systems level account of the complexity of brain functions.

Two particular findings from our research are of potential educational interest. First, in two regions of the frontal cortex associated with working memory, we found a linear correlation between neural activation while undertaking fluid analogizing and a measure of verbal IQ as determined by knowledge of irregularly pronounced English words (such as 'aisle', 'yacht', 'cello', 'syncope') (Geake & Hansen, 2005). The result is consistent with the proposition that creativity requires knowledge, the internalized knowledge that we don't know we know. It follows that as teachers, we should not be apologetic for challenging our students to learn stuff.

In a second study we compared activations associated with fluid and non-fluid analogizing with letters, numbers and geometric shapes (Geake & Hansen, 2006). We found overlapping patterns of neuronal activation between fluid and non-fluid analogizing in all formats. These results suggest that analogizing is a basic cognitive process and therefore critical for successful school performance. We also found in frontal cortical working memory areas modest correlations between non-fluid analogizing, but not fluid analogizing, and general IQ test scores, suggesting that conventional IQ tests, not to mention school assessments, might not capture abilities of fluid analogizing which underpin creative thinking (Geake & Hansen, 2006). Teachers have long suspected that IQ tests, although predictive of academic success, do not reveal all there is about a child's cognitive potential. Our findings, in supporting conjectures that the brain might develop separate working memory systems for general intelligence and fluid cognition (Blair, 2006), offer an explanation of such skepticism.

Given the pivotal role for fluid analogizing in cognition, elements of creativity through the adaptive reorganization and restructuring of novel information, we propose that there might be some potential pedagogic benefit in explicitly promoting fluid analogical thinking in formal education settings (Geake & Dodson, 2005). Consequently, we have proposed a neuro-psychological model of creative intelligence which features fluid analogizing as a central *modus operandum* (Geake & Dodson, 2005). We suggest that the pedagogic route to enhancing creative intelligence lies in fluid analogical thinking, and in our ideal classroom, students would be motivated to explore how any concept or piece of knowledge is like another, and what insights these possible analogical relationships might afford. The key variable is complexity, and all students, especially the academically gifted, should be pushed to the limits of their working memory capacity for conceptual complexity.

To return to a broader scope, it is my hope that relevant and useful professional and classroom applications of educational neuroscience will increasingly become available as we gradually come to understand more about brain function through research which answers educational questions about learning, memory, motivation, and so on. Consequently, it is my hope that in the years to come, initial teacher training, together with teacher continuing professional development programs, will include coursework components of educational neuroscience in order to create new balance-points for the perennial tensions of theory vs. practice. The longer-term objective will be to support teachers in proposing educational neuroscientific research questions to initiate the action research cycles suggested at the beginning of this chapter. The Oxford Cogntive Neuroscience Education Forum has solicited teachers' questions for neuroscientists, and it's a very long list (Geake, 2009). The pursuit of the scientifically tractable questions from such a list in the neuroscience lab might well realize more applicable results than those reviewed here. That would be the ideal outcome; teacher questions which become neuroscience questions which produce neuroscience data which is applied in the classroom and evaluated for their utility. This way we can create a scientifically rigorous and professionally informative interdiscipline of educational neuroscience.

References

Blair, C. (2006) How Similar are Fluid Cognition and General Intelligence? A developmental neuroscience perspective on fluid cognition as an aspect of human cognitive ability, *Behavioral and Brain Sciences*, 29:2, pp. 109–125.

Geake, J. G. (2004) Cognitive Neuroscience and Education: Two-way traffic or one-way street? *Westminster Studies in Education*, 27:1, pp. 87–98.

Geake, J. G. (2005) Educational Neuroscience and Neuroscientific Education: In search of a mutual middle way, *Research Intelligence*, 92, pp. 10–13.

Geake, J. G. (2008) Neuromythologies in Education, *Educational Research*, 50:2, pp. 123–133.

Geake, J. G. (2009) *The Brain at School: Educational neuroscience in the classroom* (Maidenhead, McGraw Hill-Open University Press).

Geake, J. G. & Cooper, P. W. (2003) Implications of Cognitive Neuroscience for Education, *Westminster Studies in Education*, 26:10, pp. 7–20.

Geake, J. G. & Dodson, C. S. (2005) A Neuro-psychological Model of the Creative Intelligence of Gifted Children, *Gifted & Talented International*, 20:1, pp. 4–16.

Geake, J. G. & Hansen, P. C. (2005) Neural Correlates of Intelligence as Revealed by fMRI of Fluid Analogies, *NeuroImage*, 26:2, pp. 555–564.

Geake, J. G. & Hansen, P. C. (2006) Structural and Functional Neural Correlates of High Creative Intelligence as Determined by Abilities at Fluid Analogising. Paper presented at the Society for Neuroscience Annual Meeting, Atlanta, Georgia, 17 October.

Hofstadter, D. R. (1995) *Fluid Concepts and Creative Analogies* (New York, Basic Books).

Hofstadter, D. (2001) Analogy as the Core of Cognition, in: D. Gentner, K. J. Holyoak & B. N. Kokinov (eds) *The Analogical Mind: Perspectives from cognitive science* (Cambridge, MA, MIT Press), pp. 499–538.

Kratzig, G. P. & Arbuthnott, K. D. (2006) Perceptual Learning Style and Learning Proficiency: A test of the hypothesis, *Journal of Educational Psychology*, 98:1, pp. 238–246.

Luo, Q., Perry, C., Peng, D., Jin, Z., Xu, D., Ding, G. & Xu, S. (2003) The Neural Substrate of Analogical Reasoning: An fMRI study, *Brain Research: Cognitive Brain Research*, 17, pp. 527–534.

Waterhouse, L. (2006) Multiple Intelligences, the Mozart Effect, and Emotional Intelligence: A critical review, *Educational Psychologist*, 41:4, pp. 207–225.

Wharton, C. M., Grafman, J., Flitman, S. S., Hansen, E. K., Brauner, J., Marks, A. & Honda, M. (2000) Toward Neuroanatomical Models of Analogy: A positron emission tomography study of analogical mapping, *Cognitive Psychology*, 40, pp. 173–197.

8
Brain-Science Based Cohort Studies

Hideaki Koizumi

1. Introduction

A biological perspective makes new definitions of learning and education possible. Here, learning is the process of making neuronal connections in response to external environmental stimuli, whereas education is the process of controlling or adding stimuli, and of inspiring the will to learn (Koizumi, 2000a & b; 2004). These neurological concepts of learning and education are comprehensive, covering the whole human life span; in these definitions of learning and education, the meaning of environment is everything except oneself. The cases of self-learning or self-education should also be considered, which are learning or education by self-preparation of environmental stimuli.

There are two kinds of learning: passive and active. During infancy, passive learning from the natural environment is important because newly formed neuronal connections occur only when the neurons acquire signals caused by environmental stimuli during a critical period (Koizumi, 1996, 1998). The formation of the visual cortices is a typical example of a critical period. Active learning is initiated by primitive reflexes. The reaching that appears at around three months old is a typical example of active learning, as is the beginning of locomotion-like crawling. Babies have a strong will to reach their objectives, so the reaching and locomotion help the integration of brain functions. Active learning might be an autopoietic process throughout human life.

Because development is a dynamic process with age, we must observe the developmental processes over time. Furthermore, individual diversity has to be taken into account even within the case of typical development. Therefore, a cross-sectional study at one time point is insufficient. A longitudinal study over time is necessary to elucidate the mechanisms of human development by learning and education. Also, aging is a time-dependent process. Research into both development and aging require cohort studies (Koizumi, 2006).

We have been performing various kinds of cohort studies led by Dr. Hideaki Koizumi, a Hitachi Fellow and the Director of the Research and Development Division of Brain-Science & Society, RISTEX/JST. The largest one is the birth cohort study, supported by a number of universities and national laboratories, which involved approximately 200 researchers who worked at 3 local areas in Japan over five years (Japan Children's Study [JCS], to be mentioned later). The purpose of this study was to elucidate practical operational methods for future large-size birth cohort studies; in this study, we especially included observations of infants and parents/care-givers by child-neurologists or developmental psychologists as well as inquiry-based investigations including socioeconomic factors. Namely, we prepared 5 standardized observation rooms which were equipped

Educational Neuroscience, First Edition. Edited by Kathryn E. Patten and Stephen R. Campbell.
Chapters © 2011 The Authors. Book compilation © 2011 Educational Philosophy and Theory/Blackwell Publishing Ltd.
Published 2011 by Blackwell Publishing Ltd.

with sound isolation from outside, unidirectional transparent observation windows, quantitative behavioral monitoring systems including motion capture, and a large video screen to present psychological tasks. Batteries of instruments for exposure and outcome were very carefully studied and planned in advance. We are still studying the statistics from the huge amount of results, but early results are being published (Yamagata (with JCS Group), 2009; Yamagata, 2010).

In parallel with the aforementioned strict birth cohort studies, we have been running six other different cohort studies. These studies are distributed from longitudinal birth cohort studies to simple follow-up studies combined with cross-sectional studies. These 7 studies related with cohort study methods will be mentioned below.

2. Why are Cohort Studies Important?

Human cohort studies based on the concept of 'brain-science and education' will likely have three major sets of implications.

1. Human cohort studies based on brain science are expected to produce scientific evidence that will contribute to policy-making, especially on education and related issues that pose serious problems for modern human society. For example, we might uncover implications for policy on childcare, school education, or aging.
2. We will be able to assess the potential effects of new technologies on babies, children and adolescents. For example, because humans have only recently had experience with electronic information technology, we have little information on whether such technology affects the human brain and mind. If it does have effects, we need to find out what they are.
3. Human cohort studies will allow us to test hypotheses drawn from animal and genetic case studies to see if they actually apply to people. The results of animal studies can neither conclusively prove nor disprove the validity of these hypotheses. Although a number of recent animal studies have indicated links between the behavior and expression of particular genes (Fish *et al.*, 2004; Kaffman & Meaney, 2007), we do not know whether these findings have implications for human development. We should take note that the micro-anatomical structure of the human prefrontal cortex cannot be found in rodents (Wise, 2008), and be careful in applying conclusions drawn from rodent-based studies to humans.

3. Examples of Cohort Studies Based on Brain Science

1. 'A longitudinal study of twins in infancy and childhood ("Tokyo Twin Cohort Project: ToTCoP")' directed by Professor Juko Ando, Faculty of Letters, Keio University (Ando *et al.*, 2006; Ando & Ozaki, 2009; Ando *et al.*, 2009).

Human behavioral and psychological development is affected by environmental and genetic factors. The 'twin method' can reveal the relative contribution and interaction of each factor. This study constructs a population-based twin registry in the Tokyo metropolitan area. Over 1,500 pairs of newborn twins were recruited, and longitudinal studies were conducted for five years through mailed questionnaires, individual interviews, near-infrared spectroscopic (NIRS) imaging and other methodologies.

The purpose of this study is to clarify the developmental effects of genetic and shared/non-shared environmental factors on temperament, motor skills, cognitive-linguistic abilities and other behavioral traits at an early stage of human development. For example, it was clarified that a factor for the increase of mothers' rearing stress is mainly due to genetic factors in 18-month-old infants, but due to environmental factors in 24-month-old infants. The development of reading ability of Kana characters is closely related to environmental factors, but not to genetic factors. These results are expected to provide basic information useful in childcare and the education of children. A number of results from this 'Tokyo Twin Cohort Project: ToTCoP' will be published elsewhere soon.

2. 'A cohort study of autism spectrum disorders: A multidisciplinary approach to the exploration of social origin in atypical and typical development' directed by Dr Yuko Kamio, Division Head of the National Institute of Mental Health, National Center of Neurology and Psychiatry (Kamio, 2007; Kamio *et al.* 2007; Koyama *et al.*, 2009).

Autism spectrum disorders (ASDs) are developmental disorders characterized by social deficits with varying manifestations based on gene-brain-behavior relations, although the causal factors have not yet been identified. This cohort study aims to elucidate the developmental trajectory in ASDs and in typical development, and to identify the earliest signs of ASDs and their social origin in typical development. Our final goal is to establish a prospective database of behavioral development and the corresponding neural networks to understand the pathogenesis and variability in manifestations. The outcomes of this project could contribute to early detection and intervention for children with ASDs. Also, understanding the variations of social development could provide us with solutions for the current problems in school settings in Japan. Since this program will end in 2010, preliminary results of this cohort study will be available to the public soon.

3. 'Cohort studies of higher brain function of normal elders and children with learning disabilities' directed by Professor Ryuta Kawashima, New Industry Hatchery Center, Tohoku University (Kawashima & Koizumi, 2003; Kawashima *et al.*, 2005).

A major goal of this study is to create a hopeful future by overcoming the difficulties observed in a society with fewer children and an increasing number of elderly persons from the viewpoint of brain science. The specific research topics include R&D (research and development) for anti-aging methods to maintain and improve brain functions in elderly persons, and R&D for intervention methods to develop healthy brain functions in children with learning disabilities.

Two cohort studies for each group of subjects are being performed. One is a comprehensive assessment of brain function for each group from the viewpoints of psychology and neuroscience; in this study, special attention will be paid to analyzing relationships between the subject's daily life activities and the higher order cognitive functions of the brain. Another area of interest is active cohort studies with intervention for those people by randomized controlled trials. The intervention programs to be used in the active cohort studies will be provided from the results of the preceding cohort studies. In parallel with these cohort studies, simple interventions applied at several elderly care facilities have been proving the effectiveness of the Learning Therapy. Over 1,000 elderly care institutions in Japan are now using this Learning Therapy method (Uchida *et al.*, 2008).

4. 'Cohort studies on language acquisition, brain development and language education' directed by Professor Hiroko Hagiwara, Tokyo Metropolitan University (Hagiwara & Soshi, 2007).

In modern human society, in which the daily exchange of information is accelerated on a worldwide scale, it is a commonly shared view that improving communication skills is a key to success in life. This has led to a growing social interest in early education especially for foreign languages. The introduction of English as a compulsory course at the elementary school level is, in fact, under consideration by the Japanese Ministry of Education, Sports, Culture, Science and Technology (MEXT). Although acquisition of a second language from early childhood is not undesirable, our main concern is whether it has negative effects on the normal course of language development in one's native tongue. At present, there is no scientific data available on the relationship between language acquisition (both the first and second) and brain maturation. Neither longitudinal investigations nor cross-linguistic studies exist which take into consideration how language functions are acquired as the brain develops.

The objectives of this study are to 1) investigate the mechanisms of first (L1) and second language (L2) acquisition in relation to cerebral specialization and functional plasticity in the brain, 2) identify 'sensitive period(s)' for second language learning and 3) propose a cognitive neuroscience-based guideline for second language learning and education, especially for English, including the optimal ages and conditions surrounding it. The cohort studies are conducted on three types of populations: 1) English learners of Japanese, 2) Japanese learners of foreign languages in Japan and 3) native speakers of Japanese aged 2 to 5. Each study lasted about four years until 2009. Results will be published after completing the huge data sampling process. Apart from this JST program, Noriaki Yusa (also a member of this project group) has co-authored a paper related to second language acquisition (Sakai *et al.*, 2008).

5. 'Development of novel biomedical tools for student mental health' project directed by Professor Kazuhito Rokutan, Institute of Health Biosciences, The University of Tokushima Graduate School (Kawai *et al.*, 2007a & b).

Social and environmental stressors profoundly affect the normal development of children's minds and, as a consequence, mental disorders are one of the most serious problems addressed in universities worldwide. An urgent need for society is the establishment of a simple new biomental tool to objectively assess stress response and improve the quality of life of students and younger children. High throughput DNA microarray analysis of gene expression has a potential advantage in studying complex stress response.

An original DNA array specifically designed to measure the mRNA levels for stress-related genes in peripheral blood leukocytes proved successful in detecting abnormal gene expression profiles that are closely related to mental disorders (Kawai *et al.*, 2007a);

in this study, after the collection of questionnaires from freshmen, the mental state of volunteers at our university (more than 200 students) is being studied. To detect biological risk factors for mental disorders, high-throughput analyses, measurement of stress-related molecules in saliva and evaluation of brain function by optical topography are systematically performed.

Using these biological approaches, combined with commonly used questionnaires, both environmental and biological risk factors for mental disorders in students will be identified. A new microarray-based biomedical tool will be established for student mental health (Saiki *et al.*, 2008; Kawai *et al.*, 2007b). Furthermore, this study might be able to elucidate the mechanism of depression by comparing gene expressions between clinical cases and depression moods in ordinary people.

6. 'Cohort study with functional neuroimaging on motivation of learning and learning efficiency' directed by Professor Yasuyoshi Watanabe, Osaka City University School of Medicine.

This study explores the brain mechanisms of the motivation for learning, leading to proposals for, or the development of, high-efficiency learning by maintaining high motivation and lowering fatigue during the learning process. Motivation of learning or the will to learn is related to interest, creativity and reward, but is also influenced by the extent of fatigue. Because the molecular and neural mechanisms of fatigue, especially the quantification of fatigue, have been investigated at this laboratory, it is a logical extension to proceed to a study of the motivation of children with newly developed tasks, which might be in inverse proportion to the extent of fatigue, and to follow up with them as a cohort study to correlate the extent of learning motivation with that of fatigue in the learning situation. Mainly, it was planned to perform functional MRI studies to reveal the neural basis of motivation, which has not yet been well studied, while simultaneously measuring the extent of fatigue of the volunteers (adults and children during the tasks of motivational learning). This study also includes children with learning disturbances or difficulties, in which their genetic and environmental factors will be analyzed to evaluate whether their problem is in the mechanisms of motivation or something else.

Recently, the neurological basis of motivation was partly elucidated by this project. Social rewards as well as monetary reward activate the striatum in the brain (Izuma, Saito & Sadato, 2008). Furthermore, it was found that academic rewards also activate the human striatum (Mizuno *et al.*, 2008). These results will greatly contribute to understanding motivation in learning and social activities.

7. In our program of 'Brain-Science & Society', a preparation project for future large-scale developmental cohort studies is ongoing. This project, the 'Japan Children's Study (JCS)', is directed by Professor Zentaro Yamagata, Graduate School of Medicine and Engineering, University of Yamanashi, and supported in direction matters by Professor Norihiro Sadato of the National Institute for Physiological Sciences, Professor Tokie Anmei, Graduate School of Comprehensive Human Sciences, Tsukuba University, and Dr Tadahiko Maeda, The Institute of Statistical Mathematics. This directing group positions the JCS as follows:

The objectives of the JCS are to elucidate the developmental mechanisms behind 'Sociability' or 'Social abilities', and to identify the factors that make a nurturing environment suitable or unsuitable for babies and children. The mechanisms of social development, particularly the biological aspects, are still largely unknown. Hence, the JCS will be led by a well-equipped laboratory with behavioral observation, neuron-imaging techniques and statistical analysis. The following cohort studies will test the findings/hypotheses regarding the social development from the preceding laboratory studies at the population level. All components of the JCS are systematically controlled by the center in the R&D Area, 'Brain-Science & Society', within RISTEX. There are two major organizations in the JCS: one is the laboratory and the other combines the three regional groups that prepare the cohort studies.

The purpose of the laboratory is to elucidate the developmental mechanisms of sociability and its neural representation using experimental settings. Some of the laboratories are at Kyoto University (infant laboratory for behavioral experiments), the National Institute for Physiological Sciences (for neuron-imaging studies for sociability) and Tottori University (for evaluation of peer relationships). In the infant laboratory, the mental processes of sociability, from the precursors of the more complicated functions such as morals and empathy, will be systematically analyzed and quantified using rigid experimental control. The neural substrates of these processes are being depicted and analyzed using functional MRI. The results will be extended to the application of near infrared spectroscopic (NIRS) imaging for infants. Regarding peer relationships, a novel method of sociometry based on the observation of the interaction among peers in an experimental setup will be developed.

The purpose of the regional infant cohort studies is to evaluate the progress of sociability with the chronological interaction of other factors. The factors are categorized as individual and environmental. Individual contains 1) temperament, 2) mental function other than sociability and 3) medical problems (perinatal or later). Environmental includes 1) family, 2) peers and 3) social environment. These factors and sociability will be quantified using direct observation, psychological measurement and a structured questionnaire. Also, a neurologist's observation with diagnostic techniques should contribute to the clarification of neurological development.

Several types of preliminary cohorts will be launched at the same time: an infant cohort starting with a 4-month-old baby, a preschool cohort starting with 5-year-old children and so on. The results obtained from these preliminary cohort studies with around 500 children over several years will provide invaluable information for further large-scale cohort studies.

4. Conclusion

We should recall the emergence of the concept of environmental assessment in the 1980s, when people became aware of the importance of assessing the impact of science and technology. Typical examples are the ozone hole and global warming, where the natural environment is affected by human artifacts produced by civilization. People are aware of environmental issues but only from a physical viewpoint. Drastic environmental changes in civilized society might also be the result of metaphysical problems. A flood of

information, virtual media, individualism and the pursuit of efficiency might be transforming our brain and its functions. An environmental assessment from the metaphysical
aspect could be essential to providing an appropriate environment for future generations
(Koizumi, 1996, 2000b). These critical issues, and others noted above, such as the affects
of an aging population, child development, impact of technologies, mental disabilities,
second language acquisition, and so on, can only be adequately addressed by undertaking brain-science based cohort studies.

References

Ando, J., Nonaka, K. *et al.* (2006) The Tokyo Twin Cohort Project: Overview and initial findings,
  Twin Research and Human Genetics, 9, pp. 817–826.
Ando, J. & Ozaki, K. (2009) Direction of Causation Between Shared and Non-Shared Environ-
  mental Factors, *Behavioral Genetics*, 39, pp. 321–336.
Ando, J., Kakihana, S. *et al.* (2009) Cognitive Factors Influencing Early Kana Literacy Acquisi-
  tion: A behavioral genetic approach. Paper presented to the Society for Research in Child
  Development 2009 Biennial Meeting. Colorado, USA, 2–4 April.
Fish, E. W., Shahrokh, D. *et al.* (2004) Epigenetic Programming of Stress Responses
  through Variations in Maternal Care, *Annals of the New York Academy of Sciences*, 1036,
  pp. 167–180.
Hagiwara, H. & Soshi, T. (2007) A Topographical Study on the Event-Related Potential Correlate
  of Scrambled Word Order in Japanese Complex Sentences, *Journal of Cognitive Neuroscience*,
  19, pp. 175–193.
Izuma, K., Saito, D. N. & Sadato, N. (2008) Processing of Social and Monetary Rewards in the
  Human Striatum, *Neuron*, 58, pp. 284–294.
Kamio, Y. (2007) Early Detection of and Diagnostic Tools for Asperger's Disorder, *Nippon Rinsho*,
  65, pp. 477–480. (in Japanese)
Kamio, Y., Robins, D. *et al.* (2007) Atypical Lexical/Semantic Processing in High-Functioning
  Autism Spectrum Disorders Without Early Language Delay, *Journal of Autism and Develop-
  mental Disorders*, 37, pp. 1116–1122.
Kaffman, A. & Meaney, M. J. (2007) Neurodevelopmental Sequelae of Postnatal Maternal Care
  in Rodents: Clinical and research implication of molecular insights, *Journal of Child
  Psychiatry*, 48, pp. 224–244.
Kawai, T., Morita, K. *et al.* (2007a) Gene Expression Signature in Peripheral Blood Cells from
  Medical Students Exposed to Chronic Psychological Stress, *Biological Psychology*, 76, pp.
  147–155.
Kawai, T., Morita, K. *et al.* (2007b) Physical Exercise-Associated Gene Expression Signatures in
  Peripheral Blood, *Clinical Journal of Sport Medicine*, 17, pp. 375–383.
Kawashima, R. & Koizumi, H. (2003) *Learning Therapy* (Sendai, Tohoku University Press).
Kawashima, R., Okita, K. *et al.* (2005) Reading aloud and arithmetic calculation improve frontal
  function of people with dementia, *Journal of Gerontology, Series A, Biological Sciences and
  Medical Sciences*, 60A, pp. 380–384.
Koizumi, H. (1996) The Importance of Considering the Brain in Environmental Science, in:
  H. Koizumi (ed.), *Environmental Measurement and Analysis* (Tokyo, Japan Science and
  Technology Corporation [JST]), pp. 128–132.
Koizumi, H. (1998) A Practical Approach to Trans-Disciplinary Studies for the 21st Century,
  Journal of Seizon and Life Sciences, 9, pp. 5–24.
Koizumi, H. (2000a) Nurturing the Brain: The science of learning and education, *Science Journal:
  Kagaku*, 70, pp. 878–884 (in Japanese)
Koizumi, H. (ed.) (2000b) *Developing the Brain: The science of learning and education* (Tokyo. Japan
  Science and Technology Agency [JST])

Koizumi, H. (2004) The Concept of 'Developing the Brain': A new natural science for learning and education, *Brain & Development*, 26, pp. 434–441.

Koizumi, H. (2006) The Japan Children's Study (JCS): A Large Scale Cohort Study along with Child Development, *Science Journal: Kagaku*, 76, pp. 419–425 (in Japanese).

Koyama, T., Kamio, Y. *et al.* (2009) Sex differences in WISC-III Profiles of Children with High-Functioning Pervasive Developmental Disorders, *Journal of Autism and Developmental Disorders*, 39, pp. 135–141.

Mizuno, K., Tanaka, M. *et al.* (2008) The Neural Basis of Academic Achievement Motivation, *Neuroimage*, 42, pp. 369–378.

Saiki, T., Kawai, T. *et al.* (2008) Identification of Marker Genes for Differential Diagnosis of Chronic Fatigue Syndrome, *Molecular Medicine*, 14, pp. 599–607.

Sakai, K.L., Nauchi, A., Tatsuno, Y., Hirano, K., Muraishi, Y., Kimura, M., Bostwick, M. & Yusa, N. (2008) Distinct Roles of Left Inferior Frontal Regions that Explain Individual Differences in Second Language Acquisition, *Human Brain Mapping*, 30, pp. 2440–2452.

Uchida, S., Kawashima, R. *et al.* (2008) Reading and Solving Arithmetic Problems Improve Cognitive Functions of Normal Aged People. A randomized controlled study, *Age*, 30, pp. 21–29.

Wise, W. P. (2008) Forward Frontal Fields: Phylogeny and fundamental function, *Trends in Neuroscience*, 31, pp. 599–608.

Yamagata, Z. with JCS Group (2009) *R&I Project: Identification of Factors Affecting Cognitive and Behavioral Development of Children in Japan Based on a Cohort Study* (Tokyo, Japan Science and Technology Agency (in Japanese).

Yamagata, Z. (ed.) (2010) Japan Children's Study 2004–2009; Cohort study of early childhood, *Journal of Epidemiology*, 20, Supplement II, pp. 397–504.

9

Directions for Mind, Brain, and Education: Methods, Models, and Morality

ZACHARY STEIN & KURT W. FISCHER

The emerging field of Mind, Brain, and Education (MBE) is growing fast, both as a field and as a movement, marked by the founding of the International Mind, Brain, and Education Society (IMBES) in 2004 and its journal *Mind, Brain, and Education* in 2007. MBE makes possible the coalescence between disparate fields of research and various arenas of practice. As in any complex interdisciplinary endeavor, MBE produces many open questions, both empirical and theoretical, with more opening every day.

Over the past decades, MBE has been making great strides that include formal steps to officially mark a new field of research and practice (Fischer, Immordino-Yang & Waber, 2007). Alongside the Society and Journal, there are graduate-level programs in MBE springing up around the globe in places like Cambridge (England), China, and Texas (at Arlington/Dallas). The first such graduate-level academic program was started at the Harvard University Graduate School of Education in 1999. This masters program is characterized by a year-long core course in cognitive development, education, and the brain (Blake & Gardner, 2007). The training of a new generation of educational researchers and practitioners has begun. The field is poised to usher in a new era in both the science of learning and scientifically based educational reforms. Of course, any movement this exciting and potentially important is rife with complex considerations.

In this chapter we organize a set of important considerations around three themes: *methods, models,* and *morality.* Methods involve issues of quality control and interdisciplinary synthesis. The emerging field of MBE will be best served by a methodological symbiosis between theory, research, and practice. The idea of a *research school* (Hinton & Fischer, 2008; Schwartz & Gerlach, 2011) embodies the methodological innovations we have in mind and also grounds our reflections about models and morality. Models concern the types of theories that best suit the problem-focused interdisciplinarity that characterizes MBE as a field. We suggest that comprehensive models of human development that span multiple levels of analysis are the most desirable and that their validity is best determined through broad *pragmatic* criteria. In MBE, models must prove their worth by generating usable knowledge. Regarding morality, MBE methods and models must ultimately be put to *good* use, entailing that the moral issues at stake are placed openly on the table.

Educational Neuroscience, First Edition. Edited by Kathryn E. Patten and Stephen R. Campbell.
Chapters © 2011 The Authors. Book compilation © 2011 Educational Philosophy and Theory/Blackwell Publishing Ltd.
Published 2011 by Blackwell Publishing Ltd.

The goal of this chapter is to trace *directions* for MBE, not to forecast its development. Suggesting directions for MBE goes beyond merely outlining what is possible or probable to considering what is preferable. In this chapter we aim to contribute to informing and shaping this young field.

Methods: Problem-Focused Methodological Pluralism

Mind, Brain, and Education is a problem-focused interdisciplinary field that seeks to bring together biological, psychological, and educational perspectives, with the express intention of improving educational practices. Advancing a field this complex entails certain methodological innovations. We see a set of unique quality control issues shaping the direction of these innovations. Importantly, the methodological innovations that MBE requires frame the discussions about *models* and *morality*.

Being problem-focused and interdisciplinary, MBE must meet unique demands for quality control. Key directions for methodological innovation stem from two different problem areas and converge on an emergent mode of knowledge production, which we provisionally label as *problem-focused methodological pluralism* (Dawson, Fischer & Stein, 2006) and see best exemplified in terms of 'research school' collaborations. On the one hand, researchers and educators must facilitate the integration of diverse methods from different disciplines. It can be difficult to integrate and evaluate findings generated by radically different methods, such as methods in behavioral genetics geared towards heritability versus methods in psychology geared towards conceptual change. There are issues about how to bridge different *levels of analysis* and different *basic viewpoints* (Stein, Connell & Gardner, 2008). On the other hand, MBE demands *applied* research that directly addresses the needs of educators and students—which raises issues of how educators' expertise can shape research so that it can fruitfully create usable knowledge. Here we face the challenge of bringing complex interdisciplinary research into the worlds of educational policy and practice, which raises a host of practical concerns like relevance and effectiveness. These two broad problem areas—integration of diverse methods and connections to educational application (roughly aligned with the classical division of *theory* and *practice*)—point in the direction of a symbiosis between researchers and educators in the pursuit of usable knowledge.

Issues of quality control are generally agreed to be paramount in interdisciplinary work, often leading to arguments about validity and evidence (Gibbons *et al.*, 1994; Klein, 1990). While disciplines have their own internal standards of quality control, interdisciplinary endeavors outstrip the practices of specific disciplines, raising new problems concerning what qualifies as valid and valuable work. MBE is no exception. Questions about what key *desiderata* might be for efforts in MBE stem partly from the applied nature of the field and partly from the wide range of disciplinary and practical concerns the field subsumes (Fischer, Immordino-Yang & Waber, 2007).

With so many intersecting standards and goals, the criteria for what constitutes quality work in MBE are *more rigorous* than those found in single disciplines and in interdisciplinary fields that do not involve application. Specifically, depending upon the type of work, efforts in MBE must be scientifically valid, educationally relevant, and educationally effective, at least sometimes. These are demands that have faced educational research since William James (1899) published his *Talks to Teachers* at the turn of the last century, and the insight has been echoed by Dewey (1929) and Piaget (1965). The demanding nature of educational research as *applied* is not new.

What is new is the complex interdisciplinary matrix comprising MBE and the variety of pressing educational issues to be faced. The interdisciplinary matrix raises concerns about the scientific *validity* of claims made in the field. Interdisciplinary validity claims are complex epistemologically insofar as they implicate findings and methods from different *levels of analysis* (Stein, Connell & Gardner, 2008). For example, findings describing how genetic predispositions affect anatomical features at the level of the neuron can be related to findings from brain-imaging studies (such as functional magnetic resonance imaging, called fMRI) and ultimately to behavior on academic tasks. But each of these findings deals with phenomena at very different *levels of analysis*: genetic, neuronal, functional brain organization, behavior, and complex activities on academic tasks. Generally, levels-of-analysis issues arise when we attempt to bring findings and methods together that deal with phenomena of different scale and scope—spatially, temporally, or in terms of complexity. These kinds of issues are ubiquitous in MBE.

Orthogonal to level of analysis are issues having to do with differences of *basic viewpoint* (Stein, Connell & Gardner, 2008). Analyzing a student's classroom behaviors in terms of cognitive structures and then in terms of motivation and emotion involves taking up not different levels of analysis but different *basic viewpoints* on the same behavior at the same level of analysis. Levels of analysis are differentiated in terms of the scale, scope, and complexity of the phenomenon being considered, so the difference between analyses of *behavior* and analyses of *brain function*s involves level of analysis. In contrast, when phenomena are at the same level, we can usefully analyze them in terms of different *basic viewpoints*, e.g. we can attempt to give an account of a classroom behavior either in terms of cognition or in terms of emotion (or some combination of the two). Likewise, the difference between *describing* the hyperactive behavior of a child and *evaluating* the worth of that behavior (e.g. 'it's unacceptable', or 'it's not anything to be alarmed about', etc.) is a difference of basic viewpoint, not of level. Generally, issues surrounding differences between basic viewpoints arise when we attempt to bridge methods and findings that presuppose different fundamental and deep-seated orientations toward what is being researched. The need to integrate and span multiple basic viewpoints pervades MBE.

A seductive temptation in building MBE is *reductionism* in analyzing phenomena that are studied at several levels of analysis or from different basic viewpoints. The tendency to offer unidimensional solutions to multidimensional problems is great—discussing a multi-level issue as if it can be reduced to one level or treating a multi-viewpoint issue as if one viewpoint is essential and the other can be omitted or neglected. For example, the press commonly presents findings from biological methods, such as genetics and neuroscience, as if they involve 'harder', more substantial, more scientific knowledge—privileged relative to psychological and cultural methods, which are marginalized as 'soft', needing to be reduced to biological 'causes'. Counteracting this tendency in MBE are the more direct connections of the behavior and culture to learning and schools, captured in the withering critiques of reductionism that deflated the hubris of logical positivism and its reductionist premises (Habermas, 1988; Piaget, 1971; Nagel, 1986; Sellars, 1963; Whitehead, 1925).

Moreover, the *problem-focused* nature of MBE precludes the undue marginalization of ostensibly 'softer' methods, such as those based on qualitative analyses. The fMRI machine is very different from the classroom. What seems like a valuable explanation in one place may seem hopelessly divorced from what is relevant in the other. Educational issues involve values in shared cultural frameworks. Different from the abstract problems of the laboratory, they necessarily enlist a wide variety of methods, levels of analysis, and basic viewpoints.

This brings us to the other problem area shaping the trajectory of methodological innovations in MBE: *the need to bring research into relation with educational practice and policy*. Both the framing of educational problems and the implementation of proposed solutions require the collaboration of educators and students with scientists. With the improvement of educational practices as a stated goal, MBE cannot be a solely scientific field. Most important educational issues are only resolvable in light of practice. If we admit this, then we must admit that MBE is unlike strictly scientific fields, because the progress of MBE is wedded to the progress of educational practices. And while this is not the place to recount the challenges facing efforts at scientifically based policy reform (Lagemann, 2000), it is clear that the standard conception in which scientists hand over their results from on high to educators in the trenches is not working. New collaborative relations are needed. Thus we suggest that progress in MBE requires a kind of symbiosis between educators and researchers capable of exercising quality control in the field— forms of cooperation that enable reciprocal feedback between educators and researchers to create usable knowledge and improve educational practice.

Importantly, this type of cooperative innovation, both institutional and methodological, places high demands on work in MBE, while it simultaneously provides a kind of practical quality control. It may be that conducting research in the context of practice is the best way to sort out the complex interdisciplinary and epistemological issues surrounding differences between levels of analysis and basic viewpoints. That is, we are suggesting that the validity of work in MBE is ultimately determined by a kind of *pragmatism*.

This is a theme we will return to when we discuss the *models* best suited to a field like MBE. The basic insight is that the purest usable knowledge entails a kind of openness toward different methodological approaches and traditions. With real tangible educational problems before us, we cannot afford to unduly marginalize methods just because they are geared to certain levels of analysis or basic viewpoints. In the laboratory, a good tactic is often to bracket certain perspectives in order to isolate factors and to clarify and simplify findings, but connecting research with educational import requires framing it more broadly, making explicit connections to the perspectives relevant to practice. In light of a symbiosis between educators and scientists a broad approach toward complex interdisciplinary problems emerges, which is best characterized as *problem-focused methodological pluralism* (Dawson, Fischer & Stein, 2006; Stein, Connell & Gardner, 2008).

One potential embodiment of such an approach is the idea of a *research school* (which extends from Dewey's [1896] university school or laboratory school), in which research-based educational innovations are experimentally implemented, learned about, re-tooled, and re-implemented by a community of educators and researchers working closely together. According to this idea, various researchers employing various methods work to collaborate in a problem-focused manner with teachers and administrators to deal with practical problems that are not pre-packaged for the laboratory, i.e. problems spanning multiple levels of analysis in implicating a variety of basic viewpoints. In these contexts, the ultimate criterion of success is the improvement of educational practice that results from an improved understanding of learning and teaching. Importantly, properly established research school collaborations can also facilitate the proper handling of issues surrounding *models* and *morality*, as we will explore below.

In addition, the establishment of communication channels, such as journals and conferences, are important. MBE requires both the creation of a strong research foundation for educational practice and at the same time, clear, rigorous, and responsible communication to a variety of audiences. Thus, in MBE more than mere scientific acumen is needed for work to qualify as valid and valuable. Broad overarching concerns about the future of educational practice serve to frame a problem-focused methodological pluralism. This general structure of the field has implications for theoretical model building, and it implicates MBE in moral issues concerning the means and ends of educational institutions generally.

Models: Broad Frameworks for the Epigenetic System in Context

The problem-focused methodological pluralism that we have sketched for MBE sets certain directions for building explanations and theoretical models. With multiple methods and diverse kinds of results, the need for broadly integrative theoretical frameworks increases. Likewise, the needs to communicate to wide audiences and to affect practice entail the articulation, however provisional, of big-picture accounts of how the field hangs together. Specifically, MBE needs theoretical models that span multiple levels of analysis and basic perspectives to offer *comprehensive explanations* that are grounded in multiple methodologies and focus on *processes of learning and development*, which are at the center of education. These need to be models of *the epigenetic system as a whole and in context*, which draw upon diverse perspectives and findings while avoiding simplistic and reductionist accounts.

The work of Jean Piaget (1971, 1972) sets an important precedent here. As his ambitious research program in *genetic epistemology* unfolded across decades, he offered a series of models involving explanatory constructs that cut across biological, psychological, and epistemological levels of analysis. The controversial specifics of Piaget's models need not concern us here (see Fischer & Bidell, 2006; Smith, 2002). What is relevant, and we think beyond dispute, is the value of his vision of *a comprehensive understanding of human developmental processes* (Rose & Fischer, 2009).

Some recent efforts in neo-Piagetian theorizing have kept Piaget's ambitions alive, updating his vision in light of recent research, while jettisoning some of his arguments that have not stood the test of time. Of particular note for MBE are efforts in *neuroconstructivism* (Karmiloff-Smith, 1992; Mareschal *et al.*, 2007), which offer models that subsume findings from neuroscience, genetics, and cognitive development. Such broad explanatory frameworks bring coherence to the unwieldy diversity of findings from diverse disciplines. These empirically grounded models reach beyond a single discipline or method and connect with a variety of audiences to frame educational issues and activities.

Fischer's *dynamic skill theory* is a related case in point (Fischer & Bidell, 2006). It aims to integrate the many influences on human behavior, counteracting the fragmentation of knowledge about human development that has come with the increasing differentiation and specialization of disciplines. By articulating constructs and methodological principles that cut across multiple levels of analysis, *dynamic skill theory* serves to frame the epigenetic system in its full complexity. The integration of methods from brain science, cognitive science, affective science, social analysis, dynamic systems modeling, qualitative structural analysis, and developmental assessment reveal human development as a dynamic process sensitive to contexts both biological and social. Research has revealed the educational relevance of this model and demonstrated that it can be operationalized in cycles of research and application in schools and other learning situations (Dawson & Stein, 2008).

Importantly, both the neuroconstructivist and dynamic skill theory frameworks involve general developmental principles that can make sense of *variability* and *individual differences*. These models frame learning disabilities, for example, in terms of the same explanatory principles that can account for typical development (for example, Schneps, Rose & Fischer, 2007). The image that emerges is one in which *unique* learning pathways unfold in terms of *ubiquitous* developmental processes. The ability of these models to account for diverse developmental trajectories sets them apart from many other models and makes them far more valuable in framing and integrating research for educational purposes. While Piaget (1970, p. 52) admitted that his overall approach was not amenable to the understanding of individual differences, some others who offer developmental models with wide integrative ambitions (e.g. Thompson, 2007; Tomasello 1999) do not seem to feel the need to address variability and difference at all. It seems that when scholars build theories, they often trade sensitivity to unique instances for broader scope and explanatory power, but that tradeoff is not necessary. As dynamic skill theory demonstrates, starting out with variability and individual differences can lead to principles that simultaneously (a) accommodate the messy reality of learning and development in real world contexts and (b) are *more* usefully abstract than typical psychological constructs (for example, the concepts and principles of dynamic systems theory).

A common critique is that such broad, abstract models cannot be disproved empirically. For example, Piaget's theory is criticized as too broad and vague. Nevertheless, it has generated a remarkable explosion of new research across many fields, much of it linked directly to educational research and practice (for example, Adey & Shayer, 1994; Griffin & Case, 1997). No doubt, such comprehensive models are not readily amenable to simple falsification. However, with pragmatic MBE criteria that emphasize both educational efficacy and scientific acumen, a model is subject to rigorous testing of both empirical predictions and practical usefulness (see Elgin, 2004). These broad models must pass the rigorous practical test of *creating usable knowledge*.

The emphasis on usable knowledge does not preclude basic research, but instead requires including it as part of the field of MBE. The kinds of broad theoretical models of human development that are needed to frame MBE could not be built without a great deal of *strictly* theoretical and empirical work, such as neural-network modeling or fMRI paradigms. The point we are making about the fundamentally *pragmatic* criteria by which MBE models prove their worth bears on the models' implications, relevance, and efficacy for educational practice. That is, in MBE key problems are ultimately posed and solved in the world of educational practice. While basic research will always *contribute* to the solving of these problems—by informing broad theoretical models, for example—only research carried out in the context of practice and application will *explicitly address* the key problems of MBE.

Thus in MBE purely theoretical work in fields such as psychology and brain science needs to be re-framed and adapted from its place of origin—as answers to abstract academic problems—in order to address the concerns of educational practice. Generally, we think broad theoretical models geared toward practice can serve to integrate and frame a wide variety of theoretical findings and methods. This is demonstrated, for example, by dynamic skill theory, which assimilates and integrates work from various fields and research topics under general principles that address individual differences and real world variability and learning (Fischer & Bidell, 2006).

In other words, a symbiosis of theory and practice is needed to exercise quality control in problem-focused interdisciplinary fields such as MBE. Models are built for a purpose, and models in MBE must do more than yield explanations. The notion of a research school is helpful in grounding what it means for a model to generate usable knowledge. In a research school, researchers and educators with different backgrounds converge on a common educational problem space, where they craft a common language to communicate, frame specific problems together, and articulate possible solutions. A broad and comprehensive model of learning processes can facilitate this kind of common language, e.g. here is what we mean by abstract thought, here is what we mean by learning disability. Moreover, a broad model can provide a common *language of evaluation* that is both scientifically rigorous and practically meaningful, illuminating the reform of teaching practices through a broad framework for assessment and evaluation. A good model will provide standards for evaluating student progress that are valid in the eyes of both researchers and practitioners. Efforts underway using dynamic skill theory demonstrate the power of such common standards (Dawson & Stein, 2008). Thus the most valuable models in MBE will be those that can offer this kind of usable knowledge, providing a comprehensive language and conceptual framework capable of helping to align researchers and practitioners and translate between theory and practice.

As educators and researchers use broad, integrative frameworks to build problem-focused models relevant to key educational questions, we need to consider how they can be put to *good* use—issues of ethics and values that reach to the heart of MBE.

Morality: The Ends and Means of MBE

Moral issues figure prominently in debates about genetics, neuroscience, and education because they relate to the nature of institutional structures and social norms (Habermas, 2003). We cannot address this entire topic, but will focus on issues involving the symbiosis between theory and practice in MBE. Changes in educational practice and the implementation of scientifically based interventions depend on moral decisions—about goals and values for education and learning and the scope and nature of actions to achieve those goals. Generally, two complementary ethical themes are important. One concerns *limiting* the scope of certain possible scientifically based interventions in order to preserve the integrity and autonomy of individuals. *Not all knowledge that can be used ought to be.* The other theme concerns the *fair distribution* of benefits accruing from improvements of practice. *Knowledge that is put to good use should be used to help everyone.*

A long tradition of moral philosophy can be traced from Habermas and Rawls back through Dewey and ultimately to Kant, which maintains that the goals of our most important public institutions should be determined through processes of public will formation that are subject to continual revision. These goals, provisionally set by the community, serve to orient the use of new knowledge, insuring that the institutionalization of values proceeds dialectically, as science serves the ends that people set for themselves. Generally, MBE needs to position itself in broader value-laden discussions about the goals of education. It must find a way to address the moral considerations that emerge as its fund of usable knowledge expands.

There is a set of uncontroversial educational goals that garner nearly universal consensus in countries with democratic and post-industrial forms of government. Putting aside debates about character education and the unforeseeable skills needed for tomorrow's economy, few people dispute the value of literacy, numeracy, and a host of other basic capabilities. As MBE makes progress in creating new ways of affecting learning and development, important questions will arise about the kinds of things that people should do to foster these capabilities. Some of the kinds of questions are already evident: Genetic screening and behavior modification through drugs are only the most obvious red flags in an ever-growing list of scientifically based techniques that can affect educational outcomes. There must be *limits* on the extent to which public goals can intrude upon people's private lives. Even projects aimed at fostering collectively held values such as education must stop short of transgressing the dignity and autonomy of individuals.

Following Habermas (1996), we believe that scientists and lawmakers cannot determine these limits alone. Ultimately, the voices of those potentially most affected by scientifically wrought changes in educational practice need to be heard. When it comes to issues of broad public concern (which pervade education) what is acceptable or not should never be decided *a priori* or in isolation from debates in the public sphere about the kinds of communities and individuals that ought to be fostered. Thus, public forums are required that promote general discourse aimed at articulating the ethical limits that should be placed on use of biological and psychological technologies in education. Experts in MBE will have an important role to play in this discourse, but they will be one voice among many. In a world where the development and dissemination of science and technology proceeds at alarming rates, discourse about self-imposed limitations on these developments needs to be accompanied by processes for codifying institutional policies and governmental laws. Building on reasoned public discourse, procedures must be established to enforce limits and principles for use of MBE tools and techniques.

Along with discourse about *limiting* the application of certain types of advances in MBE must come discourse about the *fair distribution* of potential benefits. Following Rawls (1971) we think that important issues about the *just* distribution of educational goods will become central with advances in MBE. And these issues must be debated in light of their ethical significance, i.e. from the perspective of a universalistic moral point of view. There must be discourse and debate about how to make innovations widely available across racial, socio-economic, and international divides, involving discussion of both how valuable educational goods are (both for individuals and communities) and how generally *unjust* their distribution tends to be in most nations. This discourse in the public sphere needs to address laws and procedures for codifying policies and laws, including active steps to ensure that the advances and benefits of MBE will not be limited to improving the learning of a privileged few. The best education systems in the world (for example Finland's) are characterized by their commitment to equality and fairness in the distribution of educational recourses (OECD, 2007). *Justice as fairness* should be a guiding principle in efforts toward scientifically based reformations of educational systems, making them both more equitable and more effective.

For both limitations and fair use of MBE knowledge, the broad community must participate in the debate. Scientists, philosophers, lawmakers, teachers, parents, and students must all weigh in. Too much is at stake for the conversation to be unduly truncated or limited to debates among elites. And so we return again to the need for symbiosis and dialogue among the stakeholders in this most important enterprise. A properly established research school could aid us in these efforts by mobilizing a diversity of voices to co-create educational environments. In such a research school, the iterative process in which research-based educational innovations are experimentally implemented, learned about, re-tooled, and re-implemented can remain open at all crucial junctures to the input from the people most affected—students, teachers, and parents. One format could be a series of structured town-hall meetings for a single school or school district where researchers, practitioners, students, and parents collaboratively frame the issues most essential to the betterment of their schools. Of course these miniature experiments in democracy will be rife with conflict and complexity, like democracy in general. The discourse can serve to inform institutional policies and practices, beginning a movement to create a culture concerned about both the ethics and the science of educational practice and capable of flowing upwards and outwards towards broader state and national policy.

Conclusion

The emerging field of Mind, Brain, and Education must be shaped not only by its constituent scientific disciplines but also by its applications to education and learning. The need for synthesis across scientific and practical disciplines points toward a problem-focused methodological pluralism and a catalytic symbiosis of theory, research, and practice. The goal of producing usable knowledge shapes the construction of theoretical models that provide holistic understanding, requiring comprehensive models of human learning and development that span multiple levels of analysis and multiple perspectives, instead of a narrow focus on truth in a scientific discipline. Ultimately these models must be put to *good* use, creating usable knowledge the application and dissemination of which is shaped by moral considerations based on dialogue among all those potentially affected. The search for usable knowledge in MBE requires focusing on quality control in light of educational relevance and efficacy as well as scientific validity.

References

Adey, P. S. & Shayer, M. (1994) *Really Raising Standards: Cognitive intervention and academic achievement* (London, Routledge).

Blake, P. & Gardner, H. (2007) A first course in Mind, Brain, and Education, *Mind, Brain, and Education*, 1, pp. 61–65.

Dawson, T. L., Fischer, K. W. & Stein, Z. (2006) Reconsidering Qualitative and Quantitative Research Approaches: A cognitive developmental perspective, *New Ideas in Psychology*, 24, pp. 229–239.

Dawson, T. L. & Stein, Z. (2008) Cycles of Research and Application in Education: Learning pathways for energy concepts, *Mind, Brain & Education*, 2, pp. 89–102.

Dewey, J. (1896) The University School, *University Record (University of Chicago)*, 1, pp. 417–419.

Dewey, J. (1929) *Sources of a Science of Education* (New York, Liveright).

Elgin, C. (2004) True Enough, *Philosophical Issues*, 14, pp. 113–131.

Fischer, K. W. & Bidell, T. R. (2006) Dynamic development of action and thought, in: W. Damon & R. M. Lerner (eds), *Theoretical Models of Human Development. Handbook of child psychology*, 6th edn., Vol. 1 (New York, Wiley), pp. 313–399.

Fischer, K., Immordino-Yang, M. H. & Waber, D. (2007) Toward a Grounded Synthesis of Mind, Brain, and Education for Reading Disorders, in: K. Fischer, J. H. Bernstein & M. H. Immordino-Yang (eds), *Mind, Brain, and Education in Learning Disorders* (Cambridge,: Cambridge University Press).

Gibbons, M., Limoges, C., Nowontny, H., Schwartzman, S., Scott, P. & Trow, M. (1994) *The New Production of Knowledge: The dynamics of science and research in contemporary societies* (London, SAGE Publications).

Griffin, S. & Case, R. (1997) Rethinking the Primary School Math Curriculum, *Issues in Education: Contributions from Educational Psychology*, 3:1, pp. 1–49.

Habermas, J. (1988) *On the Logic of the Social Sciences*, S. Nicholsen & J. Stark, trans. (Cambridge, MA: MIT Press).

Habermas, J. (1996) *Between Facts and Norms: Contributions to a discourse theory of law and democracy*, W. Rehg, trans. (Cambridge, MA, MIT Press).

Habermas, J. (2003) *The Future of Human Nature* (Cambridge, Polity Press).

Hinton, C. & Fischer, K. W. (2008) Research Schools: Grounding research in educational practice, *Mind, Brain, and Education*, 2:4, pp. 157–160.

James, W. (1899) *Talks to Teachers on Psychology: And to students on some of life's ideals* (New York, Holt).

Karmiloff-Smith, A. (1992) *Beyond Modularity: A developmental perspective on cognitive science* (Cambridge, MA, MIT Press).

Klein, J. T. (1990) *Interdisciplinarity: History, theory, and practice* (Detroit, MI, Wayne State University Press).

Lagemann, E. (2000) *An Elusive Science: The troubling history of educational research* (Chicago, IL, University of Chicago Press).

Nagel, T. (1986) *The View from Nowhere* (Oxford, Oxford University Press).

Mareschal, D., Johnson, M., Sirois, S., Spratling, M., Thomas, M. & Westermann, G. (2007) *Neuroconstructivism: Volumes i & ii* (Oxford, Oxford University Press).

OECD (2007) *PISA 2006: Science competencies for tomorrow's world. Vol. 1: Analysis* (Paris, OECD).

Piaget, J. (1965) *Science of Education and the Psychology of the Child* (New York, Viking Press).

Piaget, J. (1970) *Main Trends in Psychology* (New York, Harper & Row).

Piaget, J. (1971) *Biology and Knowledge* (Chicago, IL, University of Chicago Press).

Piaget, J. (1972) *The Principles of Genetic Epistemology*, W. Mays, trans. (London, Routledge & Kegan Paul).

Rawls, J. (1971) *A Theory of Justice* (Cambridge, MA, Harvard University Press).

Rose, L. T. & Fischer, K. W. (2009) Dynamic Development: a neo-Piagetian perspective, in: J. I. M. Carpendale & U. Mueller (eds), *The Cambridge Companion to Piaget* (Cambridge, Cambridge University Press), pp. 400–421.

Schneps, M. H., Rose, L. T. & Fischer, K. W. (2007) Visual Learning and the Brain: Implications for dyslexia, *Mind, Brain, and Education*, 1:3 pp. 128–139.

Schwartz, M. & Gerlach, J. (2011) The Birth of a Discipline and the Rebirth of the Laboratory School for Educational Philosophy and Theory, *Educational Philosophy and Theory*, 43: 1–2.

Sellars, W. (1963) Philosophy and the Scientific Image of Man, in: *Science, Perception, and Reality* (New York, Humanities Press).

Smith, L. (2002) Piaget's Model, in: U. Goswami (ed.), *Blackwell Handbook of Childhood Cognitive Development* (Oxford, Blackwell).

Stein, Z, Connell, M. & Gardner, H. (2008) Thoughts on Exercising Quality Control in Interdisciplinary Education: Toward an epistemologically responsible approach, *Journal of Philosophy of Education*, 42: 3–4, pp. 401–414.

Thompson, E. (2007) *Mind in Life: Biology, phenomenology, and the sciences of mind* (Cambridge, MA, Harvard University Press).

Tomasello, M. (1999) *The Cultural Origins of Human Cognition* (Cambridge, MA, Harvard University Press).

Whitehead, A. N. (1925) *Science and the Modern World* (New York, MacMillan Publishing Company).

10

The Birth of a Field and the Rebirth of the Laboratory School

Marc Schwartz & Jeanne Gerlach

Introduction

The explosion of new ideas and findings throughout the 20th century launched many new disciplines, and promising associations between these disciplines in turn gave birth to innovative fields of study. Both processes reflect an important feature of human behavior—primarily the desire and ability to organize information to increase its usefulness. This effort continues into the 21st century as new insights in human behavior and the brain portend new strategies to improve the learning sciences. Researchers and educators are attempting to define the limits and potential of new research as well as the suggestive findings it generates at the intersection of the cognitive science, neuroscience and education.

The emergent conversations between scientists and educators are unfolding in groups such as Brain, Neurosciences and Education[1] and Neuroscience in Education.[2] Also, in the last five years, the International Mind, Brain and Education Society (IMBES) emerged;[3] its first international conference was held, and numerous workshops, meetings, and institutes attempting to align the three perspectives have convened. In particular, the opportunity to enrich the conversation and support progress in the learning sciences led to a focused effort in defining a new field, MBE. Mind, Brain and Education is an attempt to create synergy amongst three dominant and distinct domains around the world (Harvard Graduate School of Education, 2007; Woo, 2007; Tillmanns, 2008). Toward that end, the effort has led to graduate degree programs, institutes, conferences, new projects, and one of particular interest to the field, the re-birth of the laboratory school.

In a new incarnation of Dewey's original idea (1907), the authors of this chapter are looking beyond the walls of any one building to find partners throughout a school system and within the community. The new vision is a network of schools, principals, teachers and policy makers; not necessarily everyone, but a critical mass of vested partners working collaboratively in what we call a Research Schools Network. The concept is an extension of the laboratory school where MBE becomes a theoretical and practical foundation for uniting three disciplines. Here in the southwest United States, educators and researchers at the University of Texas at Arlington have set out to find partners in the public schools, neighboring universities, the business community and legislature to develop and define new paradigms that will not only promote MBE as an academic field, but will inform and be informed by the community it seeks to serve. The goal of the MBE

Educational Neuroscience, First Edition. Edited by Kathryn E. Patten and Stephen R. Campbell.

initiative in Texas is to add leverage to the work of all individuals invested in the health and growth of education. Although the outcomes from this experiment will need to wait for another day, this chapter lays out the guiding principles and challenges in building a dynamic and synergistic collaboration.

The Birth of a Field

The roots of the Southwest Center for Mind Brain and Education originated in a series of conversations in the mid 1990s at Harvard University during a period President Bush proclaimed as the Decade of the Brain (Bush, 1990). The conversations began as an exploration of the intersections and boundaries between Mind, Brain, and Behavior (MBB). By the end of the 1990s, researchers at the Harvard Graduate School of Education saw the opportunity and potential in focusing on one specific behavior within the MBB dialogue—education (Fischer *et al.*, 2007). As a result, MBE became the intellectual space to consider how neuroscience, cognitive science and educational practice could inform each other (Harvard Graduate School of Education, 2007). However, despite the interest and power in these conversations, the new insights did not yet constitute a field of study.

To function as a field, MBE would need to manage several challenges. The first is to insure that individual disciplines like the cognitive sciences had delineated laws that could serve as conceptual frameworks for the 'science of mental life' (Dehaene, 2007; James, 1890). And for the first time in the history of psychology, the cognitive and neuro sciences appear to be well positioned to meet this challenge. Mental representations, a prominent feature of the mind, may serve as a cognitive-neural Rosetta Stone linking functional imagery and electrical activity with the behavioral outcomes that educators are interested in building, developing, and supporting (Szucs & Goswami, 2007). This development is a promising step in a field that must still define the paradigms for studying and solving a variety of educational challenges. In concomitant fashion, MBE will also have to take responsibility for clarifying findings and monitoring expectations in research conducted in its name.

All fields face similar challenges in identifying the compelling frameworks for understanding phenomena of interest, which in turn directs its member's attention to what should be measured, how and why it is measured, and which tools should be used to measure the 'it' under scrutiny. These considerations revolve around an additional challenge, which is to monitor the interaction between findings and conclusions. Imaging tools that have become available to brain researchers during the last two decades have been the basis of enthusiasm and hope for changes in policy, practice, and understanding. Just like the telescope that allowed scientists to see further and dream more deeply about the universe, brain imaging techniques are offering a new generation of researchers the possibility to dwell deeply upon how the brain functions and how these findings contribute to an understanding of ourselves. However, along with the promise has come caution; most notably, Bruer's (1997) warning of transitioning too quickly from the fascinating (perhaps intoxicating) fMRI pictures of our brains in action to decisions about how to teach. As a new field, MBE must build on public interest of the latest tools and its findings, while preventing unchecked, unrealistic and untenable claims from exhausting public enthusiasm for neuroscience research.

Another important challenge for the field is ethical dilemmas, such as those emerging at the frontier of genetics and education. Researchers can already envision genetic testing relevant to education. Within reach are tests that promise to correlate gene sequences with the appearance of dyslexia or ADHD (Galaburda *et al.*, 2006; Marlow *et al.*, 2001; Paquin *et al.*, 2006). However, these new tools also create important considerations such as, under what conditions will these tests be used or how will the results be interpreted (Grigorenko, 2007). What resources should be allocated to support and develop these tests, and will there be equal access to such screening tools? All these questions add to the ethical complexity that also emerges with these new technologies. These issues are important and central to the growth of many new fields and disciplines and the tools that emerge, and certainly MBE will need to take an active role in arbitrating over how the field should proceed in an ethical manner.

Overall, the potential of MBE to succeed as a field rests in its ability to generate new ways of understanding and solving educational problems. By employing the perspectives of other disciplines such as genetics or neuroscience, and through the use of their tools such as multivariate genetic analysis, neural networks, and dynamic growth modeling, researchers and educators can generate powerful insights into factors that contribute to learning (Van Geert, 1994; Plomin, Kovas & Haworth, 2007). Just as the brain benefits from feedback loops, MBE will benefit from feedback from the practitioners who can help fine-tune research agendas or redirect research through insights that come from years of experience with children and the learning processes they undergo.

As the conversations develop at the confluence of education, cognitive science and neuroscience, MBE faces one last challenge that has appeared in all emerging sciences— the ability to understand and respond to the different perspectives that members bring to the same conversation. In our young history, there have been two trajectories that have launched and moved the field: those of scientists and those of educators. There are subtle differences in either perspective that both groups must appreciate. Their histories and narratives are different and although MBE and educational neuroscience (and other similar groups) may converge in the future, they are still sensitive to the history and context that led to their current missions. Much of the neuroscience-education focus has emerged from the effort to help children facing specific disabilities or pathologies. The focus on understanding these problems directed researcher attention to particular neural pathways and processes, and to build the tools to localize and study these problems. The findings that these studies produced created a new opportunity for researchers and educators to speculate on how the brain functions in general and whether the findings would be of use to educators.

While these studies do provided an important platform for understanding brain functions, they are not a direct step to understanding an equally complex problem for educators; for example, how to better teach math or history in an inner city school with a diverse population where perhaps as much as 50% of the students speak some other language than English. In these complex and dynamic environments, the problems educators face are dramatically different than the individual pathologies studied by neuroscientists. Even though cognitive-neuroscience studies and their findings are crucial in contributing to our understanding of how the brain works, they do not directly address the complex environments in which learning is supported, stunted or halted.

Like the major paradigms that shaped the way the sciences have evolved in the past, MBE must be conscious of how one conceptual framework may pose the challenge of obscuring the way phenomena are understood from alternative frameworks. History provides rich and powerful illustrations of how competing points of view create years of conceptual struggles, as observed with the acceptance of heliocentrism, evolution, and continental drift (Kuhn, 1970; Le Grand, 1988; Mayer, 1991).

While MBE has benefited from (and been supported by) progress in the neurosciences by identifying relationships between human behavior (relevant to educational contexts) and brain processes, this new field must capitalize on and thoroughly integrate the perspective of educators. MBE must hold all three disciplines in a new balance, where its members ensure that the educational problems it wants to solve are clear and relevant, that the research methodologies address specific educational paradigms about learning, and that the goals of research, practice and policy are aligned.

Such a vision of cooperation and collaboration requires a context where people can address educational challenges in a supportive environment. The framework we are exploring is an extension of Dewey's lab school (Dewey, 1907). While there are current attempts to emulate the teaching hospitals where research, teaching, and health care meet under one roof (Harvard Graduate School of Education, 2007), our approach pursues a variation on this vision. Here at the University of Texas at Arlington we are trying to build a context and environment that brings together educators, administrators, researchers, and policy makers throughout a district. This vision, the Research Schools Network (RSN), is a network of the people and places invested in education who are seeking to ensure that research and practice can and do inform each other. The RSN provides a framework for defining new goals, roles and responsibilities.

The Rebirth of the Laboratory School: Challenge and Opportunity

Goals for education, such as empowering children or insuring that they can participate in a democratic society, embody universal values that once inspired Dewey (1907) and his laboratory school. However, in our experience, these goals are easily obscured by misunderstanding and fear, such as the concern about experimental interventions providing 'preferential' treatment to some students over others, or teachers who struggle to justify using classroom time for research if they are concerned that deviating from the curriculum will diminish their students' chances of passing state mandated tests. Entangled in this complex environment, researchers easily become estranged from the local contexts that provide the greatest opportunity to test educational models, and the community is unable to profit from the research process. These few educational obstacles have been sufficient to constrain the ability and imagination of all parties to build beneficial relationships that mutually enhance education. There are certainly other issues, and regrettably, they contribute to the appearance of the 'ivory tower' as the unfortunate metaphor suggesting too much distance between potential partners.

Despite these issues, the schoolhouse remains a central feature of every community and a critical nexus for parents, teachers, researchers and policy makers. If MBE is to succeed as a field, it must find ways to create ongoing opportunities to the extended educational community to collaborate, negotiate for common gain, and identify problems that are worthy of attention and study. While educators are the classroom experts, researchers offer expertise in identifying reliable strategies for testing ideas. In turn, administrators and policy makers create the guidelines to support research and classroom practice identified as best practices. To that end, the MBE Research Schools Network (RSN) is a practical infrastructure for making the goals, concerns and constraints of all parties transparent. To develop this partnership, we set out to meet four objectives: a) develop a clear vision; b) build trusting relationships; c) set a standard for rigorous research and scholarship; and d) promote meaningful assessment tools.

First, transparency insures that all parties can develop a shared *vision*. Because of the complexity of this objective, the RSN must build and support dynamic relationships between researchers, practitioners, administrators, and policy makers. The network strives for balance by keeping in focus the understanding that each partner is contributing unique skills, and thus any unified vision of success is the outcome of multiple perspectives. The Network strives for clarity by improving coordination between both the legislative and educational systems to better integrate the implicit goals embedded in standards, curricula and mandated tests. Leadership that projects these values provides a context for teachers to share their expertise by pointing with precision and detail to the specific problems that research needs to address and the system needs to support. The vision should be wide enough to provide opportunities to improve scholarship to support the ongoing process of learning, teaching, mentorship, and professional development. MBE can provide the tools, and through teamwork, identify the paradigms for meeting educational challenges. The research schools network is, thus, an infrastructure, more than a place. It is a system, more than an address. However, what is required are places where collaboration, mentorship and innovation can be tested, practiced and improved.

Second, the Network's vision depends on a structure that ensures an on-going conversation amongst members of the community, its schools, universities, policy institutions, and legislature. The RSN needs *trusting relationships*. This is a challenge for leadership, and leaders will need to understand how to support the demands of such a complex association. The SW Center for MBE is initiating this process through outreach that identifies and attempts to meet the needs of the community. By matching the concerns of teachers, parents and administrators with relevant researchers, it becomes possible to increase the leverage of research. Such interactions also reveal unanticipated insights and new challenges, as well as provide relevant expert input. The conversations help establish balance between parents and teachers seeking greater understanding (e.g. through lectures, workshops, etc.), researchers who wish to create greater understanding (i.e. through research), and finally, policy makers who wish to support the process through effective policy. Such interactions create a context where leaders, educators, parents, and researchers can observe the learning process under the influence of new interventions, using powerful instruments and assessment tools, and help develop a common language through common experiences.

Relationships that are successful in clarifying and meeting MBE's vision can more comfortably depend on *research* as a tool to advance its needs. Research ensures that the vision is viable. Research that empowers its members in turn motivates the Network to explore a wider mandate of relevant educational issues, such as how to develop and test strategies for creating future scholars, teachers, students, and leaders. As a dynamic system, the Network seeks to navigate between the needs of teachers and parents and the purpose of research because, ultimately, the researchers' ability to meet the needs of the community depends on their ability to improve the quality, rigor, focus of research, and applicability of results. MBE researchers need the community as much as the community needs the researchers.

The Network also needs to enable the next generation of teachers to see themselves as researchers as well as educators. Their involvement reduces their potential sense of isolation, and provides opportunities to apply their growing classroom expertise to educational problems. An extended community of scholars will help shape MBE by allowing all members to take part in the process of testing and understanding the limits and potential of the tools and theoretical models of this new field. These tools are often complex and mysterious. The findings can be difficult to interpret. The Network will need to use its graduate students to facilitate this process. As ambassadors as well as new researchers, they help define the value, purpose and direction of research. Their inter-actions with parents, teachers, and policy makers will help inform how their generation of researchers will fine-tune research agendas, and continue to narrow the distance between the ivory tower and the classroom.

Finally, any success in understanding whether the network is realizing its vision, is supporting an effective collaboration, and is ensuring that the research it conducts is relevant will depend on meaningful *assessments*. The current testing environment has left educators in a difficult position. Conversations that neglect the role and impact of assessments on the entire educational landscape risk outcomes where vested parties become committed to tests that do not measure what they think is being measured and/or what is valued. In a concomitant manner, the pedagogies teachers adopt to meet the demands of these assessments may not reflect what the educational community values.

Effective assessment helps close the loop between educational goals and the strategies that educators use to meet those goals. This is a simple statement, but a difficult loop to close. Effective assessments need to be very clear about what is being measured and use metrics that not only describe the achievement that students have achieved, but precisely what they would need to do to build upon their current understanding. Of particular interest are assessments shaped by what cognitive scientists understand about human development. The newest models of assessment emerging from MBE focus more on *how* students know versus *what* they know (Dawson & Stein, 2008). These assessment models are built around hierarchical levels of achievement, and can be used to evaluate the level of complexity embedded in student understandings. These tools offer teachers a vehicle to better understand the way their students construct knowledge, help teachers and researchers better strategize how to support this growth and aid policy makers to mandate standards that encourage this process.

The Future: MBE and the Research Schools Network

MBE as a field includes powerful tools, ideas and networks. Each of the four objectives (vision, relationship, research and assessment) embodies specific challenges that the field must address. MBE seeks to understand and respond to the natural as well legislative laws that shape the learning process. The field is seeking to identify the laws that will shape a new paradigm for researchers and educators to understand the learning process and enjoin the legislators to build laws that ensure educational success. MBE seeks to achieve its potential through collaboration, by offering compelling solutions to the challenges it faces, and providing a means to receive and respond to feedback. Feedback allows all parties to adjust their goals and strategies for action in a changing environment. Success is contingent upon being able to identify educational targets all individuals can agree upon and having ways of assessing its progress. The Research Schools Network offers a framework for realizing the MBE mission of creating useable knowledge and facilitating cross-disciplinary collaboration in biology, education and the cognitive and developmental sciences.

Notes

1. Their goal is 'to promote an understanding of neuroscience research within the educational community.' http://www.tc.umn.edu/~athe0007/BNEsig/ (American Education Research Association, 2009).
2. The Centre for Neuroscience in Education's goal 'is to establish the basic parameters of brain development in the cognitive skills critical for reading and mathematics.' http://www.educ.cam.ac.uk/centres/neuroscience/ (Centre for Neuroscience in Education, 2009).
3. The mission of the International Mind, Brain, and Education Society 'is to facilitate cross-cultural collaboration in biology, education and the cognitive and developmental sciences.' http://www.imbes.org/mission.html (IMBES, 2009).

References

American Education Research Association (2009) Brain, Neuroscience, and Education. Retrieved 15 March 2009 from http://www.tc.umn.edu/~athe0007/BNEsig/

Bruer, J. T. (1997) Education and the Brain: A bridge too far, *Educational Researcher*, 26:8, pp. 4–16.

Bush, G. (1990) Project on the Decade of the Brain: Presidential Proclamation 6158. Washington, DC. Retrieved 1 March 2009 from http://www.loc.gov/loc/brain/proclaim.html.

Centre for Neuroscience in Education (2009) Retrieved 15 March 2009 from http://www.educ.cam.ac.uk/centres/neuroscience/

Dawson, T. L. & Stein, Z. (2008) Cycles of Research and Application in Education: Learning pathways for energy concepts, *Mind, Brain, and Education*, 2:2, pp. 90–103.

Dehaene, S. (2007) A Few Steps Toward a Science of Mental Life, *Mind, Brain, and Education*, 1:1, pp. 28–47.

Dewey, J. (1907) *The School and Society: Being three lectures by John Dewey Supplemented by a Statement of the University Elementary School* (Chicago, IL, University of Chicago Press).

Fischer, K.W., Daniel, D. B., Immordino-Yang, M.H., Stern E., Battro, A. & Koizumi, H. (2007) Why Mind, Brain, and education? Why now? *Mind, Brain, and Education*, 1:1, pp. 1–2.

Galaburda, A. M., LoTurco, J., Ramus, F., Fitch, R. H. & Rosen, G. D. (2006) From Genes to Behavior in Developmental Dyslexia, *Nature Neuroscience*, 9:10, pp. 1213–1217.

Grigorenko, E. (2007) How Can Genomics Inform Education? *Mind, Brain, and Education*, 1:1, pp. 20–27.

Harvard Graduate School of Education (2007) Research Schools will Directly Link Research and Practice. Retrieved 1 March 2009 from http://www.gse.harvard.edu/news_events/features/2007/01/18_researchschools.html.

IMBES (2009) International Mind, Brain and Education Society. Retrieved 15 March from http://www.imbes.org/index.html.

James, W. (1890) *The Principles of Psychology* (New York, Holt).

Kuhn, T. (1970) *The Structure of Scientific Revolutions*, 2nd edn. (Chicago, IL, University of Chicago Press).

Le Grand, H. E. (1988) *Drifting Continents and Shifting Theories* (Cambridge, Cambridge University Press).

Marlow, A., Fischer, S., Richardson, A. J., Francks, C., Talcott, J. B., Monaco, A. P., Stein, J. F. & Cardon, L. R. (2001) Investigation of Quantitative Measures Related to Reading Disability in a Large Sample of Sib-Pairs from the UK, *Behavior Genetics*, 31:2, pp. 219–230.

Mayer, E. (1991) *One Long Argument: Charles Darwin and the genesis of modern evolutionary thought* (Cambridge, MA, Harvard University Press).

Paquin, B., Raelson, J., Van Eerdewegh, P., Little, R., Paulussen, R., Bruat, V., Lapalme, M., Hooper, J., Belouchi, A. & Keith, T. (2006) Whole Genome Association Studies for ADHD Using Samples from the Quebec Founder Population. Poster presentation to ASHG 2006: *American Society of Human Genetics* 56th Annual Meeting, New Orleans, 9–13 October.

Plomin, R., Kovas, Y. & Haworth, C. M. A. (2007) Generalist Genes: Genetic links between brain, mind, and education, *Mind, Brain, and Education*, 1:1, pp. 11–19.

Szucs, D. & Goswami, U. (2007) Educational Neuroscience: Defining a new discipline for the study of mental representations, *Mind, Brain, and Education*, 1:3, pp. 114–127.

Tillmanns, L. (2008) Academische School-Limburg. Retrieved 1 March 2009 from http://www.academischeschoollimburg.nl/.

Van Geert, P. (1994) *Dynamic Systems of Development: Change between complexity and chaos* (New York, Harvester Wheatsheaf).

Woo, N. (2007) Greetings from Korea! Announcement of the newly formed Korean Mind, Brain & Education Society (KMBES) Retrieved 1 November 2008 from http://www.imbes.org/newsletters/IMBES%20spring%20NL%20final.07.pdf.

11

Mathematics Education and Neurosciences: Towards interdisciplinary insights into the development of young children's mathematical abilities

FENNA VAN NES

Introduction

People tend to underestimate the complex developmental steps that are concealed behind the apparently simple nature of kindergarten mathematics. Fortunately, researchers are elucidating how young children's learning trajectories fundament formal mathematical thinking and learning (Clements & Sarama, 2007). Yet, what is concerning is that many studies and subsequent instructional interventions stay focused on children's numerical skills (Baroody, 2004). This overshadows young children's excellence in less trivial mathematical areas such as pattern and spatial perception, reasoning, classification and problem solving (Griffin & Case, 1997; Ness & Farenga, 2007).

The statement that 'early childhood education, in both formal and informal settings, may not be helping all children maximize their cognitive capacities' (NRC, 2005, p. 3) is augmented by neuroscientists who have expressed similar concerns in a report on numeracy and literacy of young children (OECD, 2002). Evidently, both mathematics education researchers and cognitive neuroscientists share an interest in fostering young children's cognitive development (De Lange *et al.*, 2008). This interest is the motivation behind the Mathematics Education and Neurosciences (MENS) project; our aim is to bridge education and neuroscience research to gain more in-depth insight into the development of young children's mathematical thinking and learning.

Bidirectional Collaboration

With publications increasingly advocating the need to integrate neuroscientific and educational research, (mathematics) educational researchers are becoming more aware of benefits of applying neuroscientific methods and findings to education (Campbell, 2006). Nevertheless, 'educational neuropsychology has remained a peripheral specialization within educational psychology' (Berninger & Corina, 1998, p. 344).

To contrast this more unilateral conception of interdisciplinarity, Berninger and Corina (1998) described how bidirectional collaboration between neuroscience and

Educational Neuroscience, First Edition. Edited by Kathryn E. Patten and Stephen R. Campbell.
Chapters © 2011 The Authors. Book compilation © 2011 Educational Philosophy and Theory/Blackwell Publishing Ltd.
Published 2011 by Blackwell Publishing Ltd.

educational research can enrich both educational research and the neurosciences. Neuroscientific research can help educational researchers gain insight into brain-cognition relations that underlie cognitive processes and the effects of instructional interventions. In turn, the behavioural patterns that educational researchers study before, during and after a particular instructional intervention, can contribute to understanding the nature of possible changes in brain activation that result as a function of learning. Thus, while 'neuroscience can investigate synaptic connections, but not conceptual ones' (Bennett & Hacker, 2003, p. 405), educational researchers can conduct conceptual investigations to understand what is known and to formulate new research questions. Such a dialogue characterizes the collaboration in the MENS project.

What concerns us in the mathematics education component of the project (hereafter ME) is the role of young children's spatial structuring and patterning ability in their later mathematical performance (Battista *et al.*, 1998; Mulligan, Mitchelmore & Prescott, 2006). Examples of spatial structures are finger patterns and domino-dot configurations. Studies have shown how children with mathematics difficulties often do not develop counting strategies beyond unitary counting (Pitta-Pantazi, Gray & Christou, 2004). We conjecture that kindergartners' spatial structuring ability can help to 'read off' quantities so that certain numerical procedures such as addition can, eventually, be abbreviated (Van Nes & De Lange, 2007). Hence, the aim of this ME research is to gain more insight into young children's spatial structuring ability and its role in the development of more sophisticated numerical procedures.

Regarding the neurosciences component of the project (hereafter NS), the focus lies in recent neuroscientific studies that have found that the automatization of extracting number meaning from symbols occurs at a very young age (grade 3) and that this can be used as a measure of the development of numerical skills (Szucs, Soltész, Jármi & Csépe, 2007). Participants in these studies decide whether a digit is numerically or physically larger than another digit that is presented simultaneously. The effect is analogous to the numerical Stroop paradigm [NSP] (Henik & Tzelgov, 1982) where adults typically show slower reaction times when numerical information is irrelevant in the physical comparison task. These outcomes not only suggest that the irrelevant numerical information is automatically processed, they also point to a developmental trajectory of response-organization skills and they provide more support for a link between impaired executive functioning and developmental dyscalculia (Szucs *et al.*, 2007). We contribute to this discussion by investigating the NSP in both a behavioural and EEG study with even younger children (kindergartners). As such, the objective is to augment current diagnostic and remedial practice with new insights into pre-requisites in the development of mathematical abilities.

Taken together, both the ME and the NS components of the project pay particular attention to young children's early spatial abilities in order to foster their mathematical development. The two disciplines are expected to enrich each other in two ways. First, the developmental phases in spatial structuring ability that we are distinguishing can be interpreted with respect to the degree in which these children automatically process quantities and vice versa. In turn, both outcomes can be interpreted in light of the children's standardized tests scores. A comparison of the three measures will help evaluate the validity of how we are gauging children's spatial structuring ability and automatic number processing. The second potential benefit of combining the two disciplines is that the instructional intervention that we are developing to support young children's spatial structuring ability can answer neuroscientific questions about the learning mechanisms that may be involved in the development of spatial and number sense. In turn, the mathematics educators can draw upon neuroscientific insights on the establishment of basic cognitive mechanisms to motivate the design of each activity in the instructional intervention.

Converging 'ME' and 'NS'

To gauge young children's spatial structuring ability and emerging number sense, we study the strategies that the children use to solve several spatial and numerical tasks that we present in a one-on-one interview setting. In addition, we designed a series of classroom activities to implement in an instructional intervention as a means to (a) better understand how children learn to make use of spatial structures for abbreviating numerical procedures, and to (b) highlight ways to amend the activities so that they interweave with children's level of mathematical understanding and support children's learning. This methodology adheres to a classroom-based 'design research' (Grave-meijer & Cobb, 2006). The activities are embedded in contexts that should intrinsically motivate the children to mathematize the problem (Freudenthal, 1971). The role of the teacher in this instruction-experiment is to guide children in discovering the advantages of applying spatial structures to solve the problems (Freudenthal, 1991). Meanwhile, the researcher observes and reflects on the children's behaviour as they are engaged in the activities. A video data analysis computer program is used to organize the raw video material and to support the building of a theory about the children's learning trajectories.

In contrast to the qualitative methodology of our ME research, the NS procedures are based on traditional experimental methods (Cohen, Manion & Morrison, 2007). In the NSP, participating children are presented in each trial with two numbers (ranging from 1 to 9, with each number in a pair differing in font size) that are simultaneously projected on a computer screen. Their task is to press the button on the corresponding side of the target (i.e. the physically or numerically larger number).

Before the trials begin, the researcher fits a cap on the child's head with electrodes that register brain activity. This non-invasive EEG technique informs the researcher about the onset and duration of brain signals for particular stimuli and motor and perceptual responses. ANOVAs help determine differences in the brain activation and in the reaction times and additional analyses give more insight into the nature of interference and facilitation effects in the different experimental conditions.

With one research discipline set in a classroom environment and another that is based on a laboratory setting, the collaboration between the ME and NS research rests on studying the same children. The children who participate in the ME research are part of the larger pool of children who will also participate in the NS research. In this way we hope to be able to compare children's phases of spatial structuring with the degree to which they automatically process quantities.

Bridging 'ME' with 'NS'

The researchers of the two disciplines adhere to their particular methodologies for different reasons. From the NS perspective, one advantage of quantitative results in the traditional experimental setting is that they can be aggregated and generalized (Jacobs, Kawanaka & Stigler, 1999). However, this methodology cannot adequately accommodate design research outcomes because the large amounts of qualitative data have to be reduced and transcribed to a more manageable amount before any analysis can begin (Strauss & Corbin, 1998). The time-consuming data collection and analyses also place limits on the size that qualitative studies can grow to.

From the ME perspective, Schoenfeld (2000) contests research that compares old and new methods of instruction or randomly assigning students to experimental and control groups because 'if different teachers taught the two groups of students, any differences in outcome might be attributable to differences in teaching. But even with the same teacher, there can be myriad differences' (p. 645). In addition, the outcomes of statistical analyses on such quantitative data only determine a (significant) difference in performance between the two groups. This contradicts the exploratory theory-building that is involved in trying to understand the complex processes that arise in instructional interventions. As a result, such outcomes have limited value in ME where the object of the research extends beyond establishing merely *that* an intervention works; rather, the focus lies on identifying successful behavioural patterns and means to support these.

After carefully considering such methodological differences, we decided that the interdisciplinarity of the project could not rely on trying to 'fuse' the qualitative design research and the quantitative experimental approach. Rather, the collaboration must 'bridge' the two disciplines within the constraints of the project so that the research traditions can meet and so that neither discipline will have to sacrifice a methodological 'virtue' for the sake of the other. As a result of this insight, the researchers of both disciplines have developed their own research trajectories that converge towards studies with the same participants. By involving the children who participated in both the ME and the NS studies, we can gather complementary information and interpret the outcomes against two different backgrounds.

From 'MENS' towards 'Educational Neuroscience'

The struggle with the controversies between the different methodologies of the two disciplines has brought us nearer to interdisciplinary research about young children's mathematical development. Already, the collaboration has led to a greater theoretical understanding of spatial structuring and emerging number sense, and has challenged us to 'think outside the box' for finding creative ways of approaching the investigation from both disciplines. The driving force behind bridging mathematics education and neurosciences in this project is the prospect of combining knowledge from both research trajectories to contribute to early diagnostic practice and prevention. If we succeed in developing and comparing two valid measures for the development of kindergartners' mathematical ability, we may help to foster young children's early mathematical insights and to stimulate those children who could be prone to experiencing difficulties in their mathematical development. The earlier we may grasp children's mathematical learning trajectories, the more we can anticipate and furnish a supportive instructional setting, and the more we may be able to support the children in the development of their mathematical thinking and learning.

Acknowledgements

This work is supported by the Netherlands Organization for Scientific Research (NWO), project number 051.04.050. The author of this chapter is supervised by Professor J. de Lange of the FIsme. They collaborate with T. Gebuis and Professor E. de Haan of the Helmholtz Institute.

References

Baroody, A. J. (2004) The Developmental Bases for Early Childhood Number and Operations Standards, in: D. H. Clements & J. Sarama (eds), *Engaging Young Children in Mathematics. Standards for Early Childhood Mathematics Education* (Mahwah, NJ, Lawrence Erlbaum Associates), pp. 173–220.

Battista, M. T., Clements, D. H., Arnoff, J., Battista, K. & Van Auken Borrow, C. (1998) Students' Spatial Structuring of Two-dimensional Arrays of Squares, *Journal for Research in Mathematics Education*, 29:5, pp. 503–532.

Bennett, M. R. & Hacker, P. M. S. (2003) *Philosophical Foundations of Neuroscience* (Malden, MA, Blackwell).

Berninger, V. W. & Corina, D. (1998) Making Cognitive Neuroscience Educationally Relevant: Creating bidirectional collaborations between educational psychology and cognitive neuroscience, *Educational Psychology Review*, 10:3, pp. 343–354.

Campbell, S. R. (2006) Educational Neuroscience: New horizons for research in mathematics education, in: J. Novotná, H. Moraová, M. Krátká & N. Stehlíková (eds), *Proceedings of the 30th Conference of the International Group for the Psychology of Mathematics Education*, Vol. 2 (Prague, PME), pp. 257–264.

Clements, D. H. & Sarama, J. (2007) Early Childhood Mathematics Learning, in: F. K. Lester, Jr. (ed.), *Second Handbook of Research on Mathematics Teaching and Learning* (Charlotte, NC, Information Age Publishing), pp. 461–555.

Cohen, L., Manion, L. & Morrison, K. (2007) *Research Methods in Education*, 6th edn. (New York, Routledge).

De Lange, J., van Nes, F. T., van Geert, P., Steenbeek, H., Feijs, E., Uittenboogaard, W. & Doorman, M. (2008) *Curious Minds: Bringing Early Reasoning Skills to the Fore*. Paper presented at the American Educational Research Association Annual Meeting, New York, NY.

Freudenthal, H. (1971) Geometry Between the Devil and the Deep Sea, *Educational Studies in Mathematics*, 3, pp. 413–435.

Freudenthal, H. (1991) *Revisiting Mathematics Education: China lectures* (Dordrecht, Kluwer Academic Publishers).

Gravemeijer, K. & Cobb, P. (2006) Design Research from the Learning Design Perspective, in: J. van den Akker, K. Gravemeijer, S. McKenney & N. Nieveen (eds), *Educational Design Research: The design, development and evaluation of programs, processes and products* (London, Routledge), pp. 45–85.

Griffin, S. & Case, R. (1997) Re-thinking the Primary School Math Curriculum: An approach based on cognitive science, *Issues in Education*, 3:1, pp. 1–49.

Henik, A. & Tzelgov, J. (1982) Is Three Greater Than Five: The relation between physical and semantic size in comparison tasks, *Memory and Cognition*, 10, pp. 389–395.

Jacobs, J. K., Kawanaka, T. & Stigler, J. W. (1999) Integrating Qualitative and Quantitative Approaches to the Analysis of Video Data on Classroom Teaching, *International Journal of Educational Research*, 31, pp. 717–724.

Mulligan, J. T., Mitchelmore, M. C. & Prescott, A. (2006) Integrating Concepts and Processes in Early Mathematics: The Australian Pattern and Structure Mathematics Awareness Project (PASMAP), in: J. Novotná, H. Moraová, M. Krátká & N. Stehlíková (eds), *Proceedings of the 30th Conference of the International Group for the Psychology of Mathematics Education*, Vol. 4 (Prague, PME), pp. 209–216.

National Research Council (2005) *Mathematical and Scientific Development in Early Childhood* (Washington, DC, NRC).

Ness, D. & Farenga, S. J. (2007) *Knowledge Under Construction: The importance of play in developing children's spatial and geometric thinking* (Lanham, MD, Rowman & Littlefield).

OECD (2002) *Understanding the Brain. Towards a new learning science* (Paris, OECD).

Pitta-Pantazi, D., Gray E. & Christou, C. (2004) Elementary School Students' Mental Representations of Fractions, in: M. Høines & A. Fuglestad (eds), *Proceedings of the 28th Annual Conference of the International Group for the Psychology of Mathematics Education*, Vol. 4 (Bergen, Bergen University College), pp. 41–48.

Schoenfeld, A. H. (2000) Purposes and Methods of Research in Mathematics Education, *Notices of the AMS*, 47, pp. 641–649.

Strauss, A. & Corbin, J. (1998) *Basics of Qualitative Research. Techniques and procedures for developing grounded theory* (London, Sage Publications).

Szucs, D., Soltész, F., Jármi, E & Csépe, V. (2007) The Speed of Magnitude Processing and Executive Functions in Controlled and Automatic Number Comparison in Children: An electro-encephalography study, *Behavioral and Brain Functions*, 3:23, doi:10.1186/1744-9081-3-23.

Van Nes, F. & De Lange, J. (2007) Mathematics Education and Neurosciences: Relating spatial structures to the development of spatial sense and number sense, *The Montana Mathematics Enthusiast*, 4:2, pp. 210–229.

12

Neuroscience and the Teaching of Mathematics

KERRY LEE & SWEE FONG NG

One distinction that is seldom mentioned in the neuroscience literature on mathematics is the difference between doing mathematics versus teaching or learning mathematics. The former is about how mathematical facts and processes are used to solve mathematical problems. This includes, for example, how primary or elementary school students perform arithmetic computation and quantity estimation. Examples from the secondary school years include how older students perform calculus, trigonometry, and algebraic questions. Although research on the teaching or learning of mathematics may also be concerned about arithmetic or algebraic computation, the focus is on the acquisition of such knowledge and how best to teach them.

Much of the available work in the neurosciences has focused on doing mathematics. The works of Dehaene and his colleagues (for a review, see Dehaene *et al.*, 2003), for example, examined the neuroanatomical systems responsible for the processing of different mathematical operations (addition, subtraction) and processes (exact calculation versus estimation). Their findings suggested that individuals represent numbers on a mental number line. Furthermore, mental arithmetic—subtraction and division, in particular—activate the intraparietal sulci, which is involved in quantitative processes involving the number line. These findings are important and tell us how people process mathematical information. However, they provide no direct evaluation of how one should one go about teaching arithmetic or encourage the development of number representation.

A point we want to emphasise is the importance of extending such research to the study of pedagogy. Take, for example, the teaching of mental subtraction. One activity often used by teachers is closely related to mental number lines. Imagine if young pupils are asked how many sweets are left when three sweets are taken away from five. Such questions can be solved by counting or direct retrieval when numbers are small. It is much more difficult when larger numbers are involved and algorithms have to be used, e.g. taking away 3 sweets from 500. One alternative, especially for weaker pupils, is to ask them to imagine a number line and counting backwards from 500. In order to use such mental number lines, the activity is first introduced with smaller numbers. Students then continue to construct and grow the number line. Such activities often have positive pedagogical effects. However, does such facilitation result from the development of students' number lines, children finding such activities fun and engaging, or some other process? Questions such as these are important, as they will help define the efficacious

Educational Neuroscience, First Edition. Edited by Kathryn E. Patten and Stephen R. Campbell.
Chapters © 2011 The Authors. Book compilation © 2011 Educational Philosophy and Theory/Blackwell Publishing Ltd.
Published 2011 by Blackwell Publishing Ltd.

boundaries of specific pedagogies. As discussed in the following sections, such questions are beginning to be addressed in the neuroscience literature.

The Neuroscience of Pedagogy

Though it is important to ascertain the neuroanatomical substrate of specific mathematical processes, it is equally important to examine how such processes are acquired. In the past, the task of translating findings from the laboratory to the classroom has largely been left to educators. As shown by recent works conducted by Delazer (Delazer *et al.*, 2005) and in our laboratory, neuroscience can play a significant role in pedagogical research. Delazer and her colleagues (2005), for example, investigated the cortical correlates of learning-by-strategy versus learning-by-rote. Participants were trained using one or the other method for over a week. The results showed that learning-by-rote activated the left angular gyrus, possibly reflecting the language dependent nature of the strategy. Learning-by-strategy, on the other hand, activated the precuneus, which the authors attributed to the use of visual imagery.

In our laboratory, we have conducted two studies investigating heuristics commonly used to teach algebraic word problems: the model method and symbolic algebra (Lee *et al.*, 2007). Problem solving is the core of the Singapore mathematics curriculum. Primary students (9 to 11-year-olds) are taught to use the model method to solve algebra word problems that, in many countries, are taught later in the curriculum using symbolic or formal algebra. The model method is a diagrammatic representation of a given problem (Ng & Lee, 2005; Ng & Lee, 2009). Take for example a question in which students are told, 'a cow weighs 150 kg more than a dog; a goat weighs 130 kg less than the cow; altogether the three animals weigh 410 kg'. Students are then asked to find the weight of the cow. Using the model method, students use rectangles to represent the unknown values as well as the quantitative relations presented in word problems (see Figure 1). Students find the weight of the cow by undoing the arithmetic procedures. It is not until students are in secondary school (12-year-olds +) when they are taught to solve such problems by constructing a system of equivalent algebraic linear equations.

Although the original intention in using the model method was to give students earlier access to complex word problems (Kho, 1987), access to non-algebraic heuristics may also complicate the acquisition of symbolic algebra. For some algebraic word problems, knowledge of such heuristics allows students to arrive at the correct solution without having to engage with the representational and transformational activities

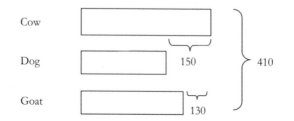

Figure 1: A model representation of an algebraic problem

associated with symbolic algebra. Some students use a mixed method strategy: combining the heuristic approach with aspects of symbolic algebra. Others construct a model drawing and then its equivalent algebraic equation before reverting to the arithmetic methods of undoing the operations. Indeed, Khng and Lee (2009) found students with poorer inhibitory abilities had difficulties using symbolic algebra even when they were directed to do so. Such students were also more likely to revert to the model method.

The potential pitfalls of introducing pre-algebraic methods are echoed by some secondary teachers' beliefs regarding the model method. Ng, Lee, Ang, and Khng (2006) reported interviews with several secondary teachers regarding the efficacy of the model method. Many of these teachers expressed reservations about the method and saw it as an obstacle to students' acquisition of symbolic algebra. One of their concerns is that the model method is childish and is non-algebraic.

How accurate are these teachers' perceptions regarding differences between the two methods? If the model method is indeed non-algebraic and poses an obstacle to the learning of symbolic algebra, it may be advisable to forego the teaching of the model method. A full evaluation of this issue will likely require a consideration of cognitive, motivational, and pedagogical issues. Because the model method is taught in all schools and has been so for over a decade, conventional programme evaluation techniques are impracticable.

In our laboratory, we conducted two studies using functional magnetic resonance imaging (fMRI) and focused on the cognitive underpinnings of the two methods. In the first study (Lee *et al.*, 2007), we asked adult participants to transform word problems into either models or equations. One of the challenges was to transform problems used in the classroom to a format suitable for use in a scanner. Classroom problems, even those used in primary schools, can be complex and require several minutes to solve. Children typically have access to pen and paper, which serve as external mnemonic aids. In some classrooms, problems are solved in a collaborative manner. Because the acquisition of images requires participants to be still and silent, it is difficult to provide such support materials in the scanner. Perhaps because of this, previous studies that examined algebraic problems have tended to focus only on numerical transformation.

Our aim was to examine differences resulting from the use of the model versus the symbolic method in solving word problems. For this reason, too narrow a focus on a specific subset of processes would not have captured the gist of our enquiry. To achieve a balance between internal and external validity, we used very simple word problems, e.g. 'John has 34 more cup cakes than Mary. How many cup cakes does John have?' To isolate processes involved in transforming word problems into models versus equations, total quantity was not presented. This discouraged participants from engaging in computation.

All participants in our study were pretested for competency in the two methods: we selected only those who were highly and similarly competent. Ensuring behavioural equivalence allowed us to infer differences in neural activation in terms of processes involved in executing the two methods rather than differences in task difficulty. Despite the lack of behavioural differences, we found differences in the degree to which the two methods activated areas associated with attentional and working memory processes. In particular, transforming word problems into algebraic representation required greater access to attentional processes than did transformation into models. Furthermore, symbolic algebra activated the caudate, which has been associated with activation of proceduralised information (Anderson *et al.*, 2004).

In a follow-up study, we investigated the next stage in problem solving: from models or equations to solution. Using models and equations derived from questions similar to those in the first study, participants were asked to solve the problems and to come up with a solution. Preliminary analyses revealed a similar pattern of findings with symbolic algebra activating areas associated with working memory and attentional processes. Findings from these two studies suggest that, for simple algebraic questions at least, differences between the methods are quantitative rather than qualitative in nature. Both methods activate similar brain areas, but symbolic algebra imposes more demands on attentional resources.

Pedagogical Implications

If symbolic algebra is indeed more demanding on attentional resources, one curricular implication is that it is best to teach the model method at the primary level and leave symbolic algebra until students are more cognitively matured. In evaluating this recommendation, it is important to note that participants in our neuroimaging studies were adults and were all similarly proficient in the two methods. Our recommendation assumes that similar differences will be found in younger learners; indeed, using symbolic algebra may require even more effort for early learners of algebra. Nonetheless, these are empirical assumptions that require further investigation.

We mentioned earlier that some secondary school students adopt a mixed-method strategy in solving algebra problems. In the following section, we discuss the intervention used to address this issue. Of relevance is that the neuroimaging findings provided some insights on why the intervention had limited success.

Students' use of a mixed method approach is of concern to educators because knowledge of symbolic algebra is critical for solving problems in higher mathematics and in the sciences. To address this issue, the Singapore Ministry of Education and the National Institute of Education jointly developed '*Algebar*', a software tool designed to (a) help students make the link between the model method and symbolic algebra and (b) acquire the direct algebraic route of problem solution. Given an algebra word problem, students who would otherwise use a mixed method approach are prompted by the software to construct equivalent algebraic equations, beginning with definitions of the variables. The inbuilt self-checking system provides instant feedback that supports the construction of algebraic equations.

In late 2006, this software was piloted in two secondary schools, involving four teachers. Participants (~13 year olds) were first taught the symbolic manipulation and transformational activities related to symbolic algebra. Working in pairs, they used *Algebar* to solve a set of algebra word problems. We videotaped the interactions between eight pairs of students. The dyadic interactions showed that many times, students drew an appropriate model representation, constructed a set of equations, and were then unsure how to proceed with constructing a set of equivalent equations that would lead them to the solution.

Focusing on the teachers, analysis of these lessons showed that they used a transmission paradigm to deliver the content. Rather than explaining the procedures necessary for constructing a system of equivalent linear equations, students were *told* how to transform equations. In the post-lesson interviews, teachers explained that their pedagogy was constrained by the limited curricular time allocated to the teaching of symbolic manipulation and transformational activities.

Why do students find constructing a system of equivalent equations difficult? One possible reason is that the pedagogy used to teach students symbolic manipulation and transformational activities was not meaningful. Because the time spent on introducing algebra was brief, students may not have sufficient time to master the procedures. This is particularly problematic if we consider our imaging findings. They show that even for simple algebra problems, symbolic algebra is resource intensive relative to the model method. One solution is to spend more time on introductory activities associated with symbolic algebra. If the procedures related to the construction of a system of equivalent linear equations are explained and are then rehearsed until they are automatized, students may find it easier to adopt the symbolic route to the solution of problems.

As part of this effort, the second author of this chapter offers professional development courses to help teachers enhance their pedagogy. The objectives of these courses are to inform teachers how they can use *Algebar* to make the link between the two methods and to help improve the teaching of symbolic manipulation and transformational activities. One specific objective is to deploy strategies that will help students reduce the working memory demands of symbolic algebra.

Conclusions

In this chapter, we focused on two issues. The first issue is related to the distinction between doing versus teaching mathematics: knowing how specific mathematical processes are implemented will not necessarily tell us how best to teach them. Second, one of the challenges in drawing useful information from the neurosciences is to bridge the divide between the laboratory and the classroom. The suggestion that neuroscience may inform the work of mathematics educators often elicits raised eyebrows from colleagues, who respond by asking whether it involves installing a fMRI machine in schools and scanning children to determine the state of their brains. More seriously, replicating the group-based characteristics of pedagogy in a platform designed for individual investigation is non-trivial. Although the efficacy of pedagogy can be investigated on a one-to-one basis, what works in an individual setting may not be effective in a classroom setting. A closely related concern relates to the context in which learning occurs. What works in a controlled laboratory environment may not do so in a classroom environment. Research in education has also emphasised the importance of discourse amongst community of learners (Brown & Campione, 1994); again something that is difficult to implement within the confines of a scanner.

Although these observations may portray a negative view of future progress, this is not our intention. Some of the identified problems are technical in nature. Recent developments in near-infrared spectroscopy and electroencephalography promise both portability and higher tolerance for movement, both of which may allow for more naturalistic examination of pedagogical strategies.

Note

Correspondence regarding this chapter should be sent to the first author at Kerry.Lee@nie.edu.sg. Readers interested in the *Algebar* programme should contact the second author at SweeFong.Ng@nie.edu.sg.

References

Anderson, J. R., Qin, Y., Stenger, V. A. & Carter, C. S. (2004) The Relationship of Three Cortical Regions to an Information-Processing Model., *Journal of Cognitive Neuroscience*, 16, pp. 637–653.

Brown, A. L. & Campione, J. C. (1994) Guided Discovery in a Community of Learners, in: K. McGilly (ed.), *Classroom Lessons: Integrating cognitive theory and classroom practice* (Cambridge, MA, MIT Press), pp. 229–272.

Dehaene, S., Piazza, M., Pinel, P. & Cohen, L. (2003) Three Parietal Circuits for Number Processing, *Cognitive Neuropsychology*, 20, pp. 487–506.

Delazer, M., Ischebeck, A., Domahs, F., Zamarian, L., Koppelstaetter, F., Siedentopf, C. M. *et al.* (2005) Learning by Strategies and Learning by Drill: Evidence from an fMRI study, *Neuroimage*, 25, pp. 838–849.

Khng, F. & Lee, K. (2009) Inhibiting interference from prior knowledge: Arithmetic intrusions in algebra word problem solving, *Learning and Individual Differences*, 19:2, pp. 262–268.

Kho, T. H. (1987) Mathematical Models for Solving Arithmetic Problems, *Proceedings of ICMI-SEACME*, 4, pp. 345–351.

Lee, K., Lim, Z. Y., Yeong, S. H. M., Ng, S. F., Venkatraman, V. & Chee, M. W. L. (2007) Strategic Differences in Algebraic Problem Solving: Neuroanatomical correlates, *Brain Research*, 1155, pp. 163–171.

Ng, S. F. & Lee, K. (2005) How Primary Five Pupils Use the Model Method to Solve Word Problems, *The Mathematics Educator*, 9, pp. 60–84.

Ng, S. F. & Lee, K. (2009) The Model Method: Singapore Children's Tool for Representing and Solving Algebraic Word Problems, *Journal for Research in Mathematics Education*, 40:3, pp. 282–313.

Ng, S. F., Lee, K., Ang, S. Y. & Khng, F. (2006) Model Method: Obstacle or bridge to learning symbolic algebra, in: W. Bokhorst-Heng, M. Osborne & K. Lee (eds), *Redesigning Pedagogies* (New York, Sense Publishers).

13
The Somatic Appraisal Model of Affect: Paradigm for Educational Neuroscience and Neuropedagogy

KATHRYN E. PATTEN

Introduction

The birth of educational neuroscience is long overdue. The transmogrification of brain research into brain-based teaching must meet its metaphorical demise. Neuromyths (Hall, 2005; Goswami, 2004; Geake, 2008) that have crept into pedagogy must be replaced by an applied neuroscience, primarily concerned with educational practice both informed by educational neuroscience *and* informing research in neuroscience and neuropsychology. This applied neuroscience may be termed neuropedagogy.

Education is still, by and large, stuck in the Hellenistic tradition of the Western world that regards the intellect as supreme and emotion as a detractor, by-product or, more recently, as a type of intelligence. While Aristotle, Descartes, and Spinoza contributed to the metamorphosis of the relationship between intellect and affect to reveal emotion as more than an irrational renegade in need of rational control; modern neuroscientists, especially the likes of Damasio, Ledoux, and Davidson, portray emotion as the phylogenetic director of embodied brain function that is distinct from cognition, and assign primacy of brain-body function to emotion.

Educational neuroscience must reflect the newly assigned importance of emotion and the concomitant role of cognition. Neuroscience and neuropsychology assert the primacy of emotion function, and do not regard emotion as subservient to cognition, nor as a type of intelligence. They present emotion as a function of brain-body interaction, a vital part of a multi-tiered phylogenetic set of neural mechanisms, evoked by both instinctive processes and learned appraisal systems. In light of neuroscientific findings, what is needed is a re-conceptualization of emotion in relation to intellect. Primarily based on Damasio's somatic marker hypothesis, but also incorporating elements of appraisal theory, this chapter presents a neuropedagogical paradigm of emotion, the Somatic Appraisal Model of Affect (SAMA).

Primacy of Emotion Function

While emotions and cognition are inter-relational, they constitute separate subsystems of the brain/body function (Davis, Hitchcock & Rosen, 1991; Gray, 1999; Panksepp, 1990, 1994). There are several arguments that support the primacy of emotion's function.

Educational Neuroscience, First Edition. Edited by Kathryn E. Patten and Stephen R. Campbell.
Chapters © 2011 The Authors. Book compilation © 2011 Educational Philosophy and Theory/Blackwell Publishing Ltd.
Published 2011 by Blackwell Publishing Ltd.

Firstly, emotions perform the first level of appraisal for incoming stimulus in brain function, thus affecting subsequent cognitive appraisals. Secondly, emotions have the capability of easily influencing bodily functions and responses; and thirdly, emotions, specifically through the functions of the amygdalae, have the ability to influence and override cognition. In addition, emotions involve many more brain systems than thoughts (Ledoux, 1986, 1996). Emotions, writes Ledoux, 'cause a mobilization and synchronization of the brain's activities' (1996, p. 300) that is functionally dissimilar to thinking that does not have significant emotional content.

As well, Gray (1991) argues that there is evidence that the limbic system constitutes a 'separable subsystem of the brain' (p. 274), and that different parts of the limbic system share a common antigen that is not present in other brain regions. He also cites a 'unity of response' in the limbic system, even though the whole system contains diverse and distinct components (Gray, 1991, p. 276).

Others avow the primacy of emotion because of its role in human behaviour. Kagan (1984) promotes the primacy of emotions when he writes that 'feelings can dominate consciousness in a way that thoughts cannot' (p. 69). Izard (1984) argues that emotion is not only primal in human development, but serves as a fundamental core of human existence that is the motivation for both cognition and behaviour. Damasio has long been an advocate of emotions as capable of influencing and enhancing decisions (1994, 2003; see also Bechara, 2004; Davidson, Jackson & Kalin, 2000). Goleman (1995) states that the intensity of emotion enhances memory and Parkinson, Totterdall, Briner, and Reynolds (1996) argue that the valence of underlying emotions or mood influences the style of thinking, the encoding of material (Ledoux, 2000; Parrott & Spackman, 2000; Phelps, 2006), as well as the storage processes. Davidson, Jackson, and Kalin (2000) state that emotion 'provides the motivation for critical action in the face of environmental incentives' (p. 890). Silvia's study notes that self-awareness of emotional experiences 'reduces egocentrism and enhances perspective taking' (2002, p. 21) and speculates that this emotion-salience can influence appraisal and regulation variables. Indubitably, emotion plays a vital role in brain/body function and influences aspects of cognition.

Need for an Educational Paradigm of Emotion

The findings of affective neuroscience reveal that understanding the neurobiology and neuropsychology of emotion and how it applies to human behaviour is not only essential to human flourishing; it is essential to overhauling pedagogy (Patten, 2004). However, while much work has been done in identifying the biological and neurological brain/body processes of emotion and the role of emotion in human behaviour, a conceptual or theoretical model of emotion that is *both* rigorously grounded in this work *and* suitable for educational researchers and practitioners is lacking. I propose a model of emotion suitable for neuropedagogy, that is, for a neuroscientifically informed and grounded pedagogy.

The Somatic Appraisal Model of Affect

Based primarily on the work of Damasio (2003), Ledoux (1996), and others (e.g. Dalgleish, 2004; Davidson, 1999; Davidson *et al.*, 2000; Forgas, 1999; Gray, 1991; Izard,

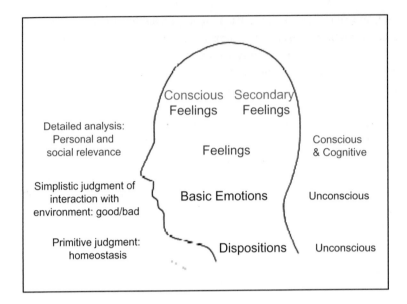

Figure 1: Somatic appraisal model of affect

1984; Panksepp, 1994), along with elements of appraisal theory (Beedie, Terry & Lane, 2005; Kagan, 1984; Lazarus, 1999; Sabini & Silver, 2005; Scherer, 2000), I propose the Somatic Appraisal Model of Affect as a new educational model for emotion.

SAMA has its roots in Damasio's somatic marker hypothesis (2003), which asserts the phylogenetic primacy of affect and presents emotion as deriving from bodily or somatic states, such as heart rate, homeostatic condition, and neuronal arousal. Provoked by incoming stimuli, bodily states are then appraised, and behaviours are evoked or inhibited based on this evaluation (Damasio *et al.*, 2000; Hinson, Jameson & Whitney, 2002). These bodily states which are prompted by signals, both neuronal and chemical (Damasio, Tranel & Damasio, 1991), from the amygdalae and other limbic regions, allow the affective brain to ascribe valences to information that, in turn, influence attention, memory, information processing, and well-being in ways that cognition does not (Maier, 1991; Meinhardt & Pekrun, 2003; Moore, Underwood & Rosenhan, 1984; Tugade & Fredrickson, 2004).

SAMA embraces an embodied philosophy of mind wherein observable behaviours and brain/body functions are objective manifestations of lived experience (Campbell, 2005a, 2005b; Ferrari, 2003). SAMA has integral elements, including clarification of the terminology of emotion necessary to articulating the body/brain process and function of emotion, and delineation of the components and facets of emotion inherent in the phylogenetic phenomena of affect (see Figure 1).

Definitions of Affect

In order to avoid disparate discourse and the resultant confusion related to what are commonly termed emotions (e.g. Cole, Martin & Dennis, 2004; Damasio, 2001; Leventhal, 1982), SAMA seeks to provide clarity regarding distinctions among the various types of what I shall hereafter refer to collectively as affect. SAMA presents three categories of affect in phylogenetic order: dispositions; basic emotions; and feelings, of which there are two types, namely conscious feelings and secondary feelings.

Dispositions, variously referred to as 'background feelings' (Damasio, 2003, p. 45), tonal emotions, internal tone (Kagan, 1984), or more generally mood, are states of being composed in the basal ganglia and limbic regions reflecting composite mind/body expressions of regulatory status in regard to homeostasis. Dispositions are largely unconscious; they register a fusion of underlying somatic states, including both chemical and visceral conditions over time, contributing to their vague nature and generalized essence. Because of their representation of the body/mind's state relative to homeostasis and basic existence, dispositions are by nature primitive and basic to survival.

The second level of affect, basic emotions, is also referred to as primary emotions (Damasio, 2003). SAMA defines basic emotions as specific and relatively consistent representations of physiological, chemical, and neural responses evoked by certain brain systems when a person perceives or recalls objects or situations. Basic emotions are prototypic and characterized by autonomic and simple judgments, positive/negative, approach/withdraw, which primarily involve the limbic system, and occur in the 'low road' (Ledoux, 1996, p. 164) of stimuli processing. This unconscious appraisal is 'quick and dirty' (Ledoux, 1996, p. 163); it evaluates incoming stimuli from the external environment in relation to survival or life preservation. Basic emotions are elemental affects in that they are largely innate (Damasio, 1999, 2003; Ledoux, 1996; Panksepp, 1990) and somewhat instinctive. Basic emotions are generally accepted to include happiness, sadness, fear, anger, disgust (Ekman, 1993), and sometimes surprise (Damasio, 2003, Porges, 2001). The essence of these emotions is present at birth, is cross-cultural (Ekman, 1993), and also exists in primates (Damasio, 2003; Ledoux, 1996). Basic emotions are 'not dependent on cognitive development for emergence nor on cognitive appraisal for their activation' (Dougherty, Abe & Izard, 1996, p. 29).

SAMA's third level of affect shall be termed feelings. Also referred to as secondary emotions, social emotions, and conscious emotions, feelings are mental recognitions of the pattern of physiological, chemical, and neural responses evoked by certain brain systems when a person perceives or recalls objects or situations. Feelings involve cortical appraisal systems and therefore have a cognitive component; they take place in the 'high road' (Ledoux, 1996, p. 164) of the brain. These cortical appraisals involve detailed analysis of incoming stimuli in relation to memory, knowledge, and a sense of self. Feelings involve two types of cognitive analysis: 1) recognition of somatic patterns or maps of body/brain conditions; and 2) the pairing of somatic maps with detailed analysis of object/event stimuli in regards to learned knowledge and experience, which include cultural norms, personal goals, and the constructed self. When the first type of appraisal occurs, when somatic maps are recognized, consciousness accompanies this recognition of somatic patterning, and these I refer to as conscious feelings. When the second type of appraisal occurs, when stimuli are evaluated in regards to the stored essence of self and prior experience, basic emotions may be transformed into what I refer to as secondary feelings. These secondary feelings are evolved basic emotions; they are cognitive in nature, they are learned, and their causal emergence may differ from person to person. Secondary feelings include such affects as shame, guilt, jealousy, and pride. Secondary feelings require analysis of stimuli in regards to personal goals, social and cultural norms, family values, a sense of self, as well as context. For example, shame may be regarded as a type of sadness combined with a touch of disgust arising from a cognitive analysis of personal performance in regards to evaluation of social and cultural norms, values, and the context in which the event occurred.

The various types of affect differ by virtue of the regions of the brain activations, levels of consciousness, and roles in influencing human function, including evoked behaviours, (Ledoux, 1996; Ohman, 1999), yet their phylogenetic composition ensures that they are interconnected physiologically, chemically, and functionally. Common sensory input, shared neuronal pathways, and interaction of neurotransmitters or chemicals that evoke or inhibit brain/body behaviour both promote and accommodate an ever-changing brain/body system. Because of the many levels of interconnection, each level of affect is capable of influencing those levels with which it has efferent and afferent communication, those levels for which it is a contributing determinant. In real life terms, a person's disposition can influence cognitive analysis and interpretations of objects and events, and ultimately, all types of affect have the ability to influence behaviour because they influence brain/body function. Because many aspects of affective responses are observable by the naked eye or through neurophysiological and psychological measurements, affect can be investigated objectively (Damasio, 1999).

Clarification of the different types is essential to a working model of affect for education. While there is not room here to address the debate between those who argue that all types of affect involve cognition, suffice it to say that for the purposes of this model, cognition in the function of affect is taken to imply that the neocortex is the most influential or evocative appraiser of stimuli. In the cognition versus affect debate, there is overwhelming support that the systems that comprise affect function primarily but not limited to the limbic region, and are distinct and capable of operating without cognitive regions of the brain (e.g. Braeutigm, Bailey & Swithenby, 2001; Damasio, 2003; Davis, Hitchcock & Rosen, 1991; Geary, 1996; Gray, 1999; Izard, 1984; Lang, Davis & Ohman, 2000; Ledoux, 1996; Panksepp, 1994; Zajonc, 1980; Zajonc, Pietromonaco & Bargh, 1982), especially for primal levels of affect, such as fear. In other words, early appraisal of incoming stimuli need not involve higher order cortical processing or conscious appraisal in order to evoke basic emotions and produce bodily responses such as changes in heart rate, sweating, muscular tension, eye blinking, chemical responses, and neuronal activation of other brain regions. The ability of the amygdalae to assess incoming stimuli for affective content and then activate other regions of the brain more rapidly and dramatically than the thinking part of the brain, the neocortex, lends credence to evidence that the amygdalae is capable of overriding the cognitive brain (Damasio, 2003; Goleman, 1995; Ledoux, 1996).

Components and Facets of SAMA

While SAMA incorporates the neuroscience or brain/body mechanisms of affect with the neuropsychology or brain/body behaviour resulting from affect, it also proposes the components and facets of affect. While there is not space here to delineate these, clarification of components and facets is essential to both researchers and practitioners;

coherence of what we are identifying, quantifying, and modifying is paramount to effectively creating and monitoring changes in pedagogy. The components of affect include elicitors, receptors, emotional states, expressions, and the emotional experience itself (Patten, 2008), and facets of affect include tone, valence, intensity, duration, context, integrity, intentionality, cognitivity, and physiology. Delineating both components and facets of affect is vital when determining the theoretical basis for a hypothesis, what entity is being examined, and how that entity will be observed and measured. Education researchers must be clear as to what it is they are hypothesizing about, what components and facets are critical to their experiments, and, ultimately, how their interpretations of affect may be applied to pedagogy.

Arenas of Cognitive Appraisal

While Damasio's SMH serves as the essential neuroscientific foundation of SAMA, his hypothesis focuses on the largely unconscious evaluation of stimuli that occurs primarily in the limbic system but utilizes other brain areas, and what is known about the processing of secondary feelings. While his hypothesis is critical to understanding the science of dispositions, basic emotions, and the emergence of feelings, it does not yet elucidate in any detail the higher behavioural levels of appraisal associated with the function of secondary feelings (see Figure 2).

It is here that appraisal theory, especially that of Lazarus' Motivational-Relational Theory of Emotions (MRTE) may have application (1994, 1999). Like the SMH, SAMA proposes the neurobiological functions of dispositions (i.e. background emotions: Damasio, 2003) and basic emotions (primary emotions: Damasio, 2003) as lower level affect where the amygdalae serve as the hub for triggering activation and coordination of other brain regions essential to affect. Basic emotions become conscious feelings when the somatic representations, or facsimiles produced by mirror neurons, are recognized by regions in the neocortex. Not only do conscious feelings involve cognition,

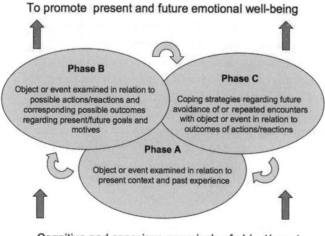

Figure 2: SAMA: Arenas of appraisal for secondary feelings

so too do secondary feelings (secondary or social emotions: Damasio, 2003), which are phylogenetically more evolved brain functions than basic emotions. While dispositions involve primitive judgments regarding homeostasis; and basic emotions involve simplistic judgments regarding external and environmental aspects of survival; feelings, both conscious and secondary, involve higher levels of detailed analysis relating to emotional well-being.

It is secondary feelings where appraisal theory may be cautiously applied. I say cautiously, since neuroscience has yet to specify the neurobiological workings of second-ary feelings. Secondary feelings, more sophisticated versions of basic emotions, both evolve and mature with neural, social, and emotional development, demanding delinea-tion to help unravel their complex nature. SAMA proposes the appraisal of secondary feelings as three interacting arenas that are collectively both cognitive and conscious. In the first appraisal arena, Phase A, the object or event is examined in relation to present context and past experience. This phase corresponds to Lazarus' primary appraisal. In the second appraisal arena, Phase B, the possible actions/reactions to the object or event and corresponding possible outcomes in relation to motives and present and future goals, are examined. This phase is similar to what Lazarus (1999) terms secondary appraisal, and can be seen to include Lazarus' action tendencies. In Phase B, possible outcomes are considered and then initiated and directed. In the third appraisal arena, Phase C, coping strategies and future avoidance strategies are formulated and considered in relation to the enacted or realized outcomes of response(s) to the stimulus.

It is in these appraisal arenas that the educationally relevant constructs of meta-emotion, consciousness of emotional contagion, and emotion regulation show promise. While there is not room here to define these constructs, the types of emotion regulation that are possible in each arena or phase have been explored (i.e. Gottman & Katz, 2002; Gross & John, 2003; Lewis & Steiben, 2004; Shipman, Zeman & Stegall, 2001; Silk, 2002).

Since the emergence of dispositions and basic emotions are to a large degree auto-nomic and unconscious, they cannot be recognized nor stopped until they become conscious feelings. However, they can be attenuated and avoided in the future through emotion regulation by recognizing their emergence triggers and enacting preventive measures related to specific objects and situations.

The critical point regarding these arenas of appraisal is that they, like brain regions and brain-body interactions, function, not as distinct units, but rather as various instruments of an orchestra playing a fugue, with a combined conscious effort to produce the best possible result, which in our case is to promote present and future emotional well-being. This complex emotive-cognitive living being that we call a human remains much of an enigma, but it is the intent of SAMA to combine what is now known from the research and theorizing of neuroscientists and neuropsychologists into a tentative whole for educators, as a foundation for neuropedagogy.

Conclusions and Educational Implications

Educational neuroscience is seen as a bridge to connect the significant differences between knowledge of neuronal function and how these functions operate and actuate in teacher/learners. As such, prescriptions of neuropedagogy both warrant and are in need of substantiation through educational neuroscience (Campbell, 2005b).

The neuroscientific body of literature that serves as a foundation for this chapter presents affect as having primacy over cognition, and having the capability of influencing, attenuating, and even subjugating cognition. Affect is revealed as paramount in significance to human mental processes such as learning, and crucial to survival, social functioning, and human well-being. Neuroscientific evidence of brain/body interaction, the primacy of emotion in human functioning, and neuropsychological data of behavioural influence elicited by affect requires the attention of educational researchers, especially educational neuroscientists, and is quite likely to eventuate in some profound changes and shifts in educational theory and practice.

To this end, SAMA is formulated and proposed as a preliminary composite framework of the basic function and facets of affect in relation to cognition within an embodied paradigm on which to build a transdisciplinary model of affect. This model identifies quintessential functions, components, and facets of affect necessary to provide a new research domain, namely educational neuroscience, with a basis on which to build a dynamic model of affect serving to challenge current pedagogy and inform and build a new praxis, called neuropedagogy.

Some quintessential questions related to the primacy of emotions in our daily, lived experiences as lifelong teacher/learners persist. We must identify and address the affective needs of teachers/learners and seek to understand how these needs impact teaching, learning, and social functioning. Many issues and avenues of research remain to be further explored in regard to affect in order to establish and legitimate neuropedagogy.

References

Bechara, A. (2004) The Role of Emotion in Decision-Making: Evidence from neurological patients with orbitofrontal damage, *Brain and Cognition*, 55, pp. 30–40.

Beedie, C. J., Terry, P. C. & Lane, A. M. (2005) Distinctions between Emotion and Mood, *Cognition and Emotion*, 19:6, 847–878.

Braeutigm, S., Bailey, A. J. & Swithenby, S. J. (2001) Task-Dependent Early Latency (30–60 ms) Visual Processing of Human Faces and Other Objects, *Cognitive Neuroscience and Psychology*, 12:7, pp. 1531–1536.

Campbell, S. R. (2005a) Specification and Rationale for Establishing an Educational Neuroscience Laboratory. Paper Presented at the American Educational Research Association Conference, Montreal, PQ.

Campbell, S. R. (2005b) Educational Neuroscience: Keeping learners in mind. Paper Presented at the American Educational Research Association Conference, Montreal, PQ.

Cole, P., Martin, S. & Dennis, T. (2004) Emotion Regulation as Scientific Construct: Methodological challenges and directions for child development research, *Child Development*, 75:2, pp. 317–333.

Dalgleish, T. (2004) The Emotional Brain, *Nature Reviews Neuroscience*, 54, pp. 582–589.

Damasio, A. R. (1994) *Descartes' Error: Emotion, reason and the human brain* (New York, Avon Books).

Damasio, A. R. (1999) *The Feeling of What Happens: Body and emotion in the making of consciousness* (New York, Houghton Mifflin Harcourt).

Damasio, A. R. (2001) Emotion and the Human Brain, in: A. R. Damasio, A. Harrington, J. Kagan, B. S. McEwen & H. Moss (eds), *Unity of Knowledge: The convergence of natural and human science* (New York, New York Academy of Sciences), pp. 101–106.

Damasio, A. R. (2003) *Looking for Spinoza: Joy, sorrow, and the feeling brain* (New York, Harcourt).

Damasio, A. R., Grabowski, T. J., Bechara, A., Damasio, H., Ponto, L. L. B. & Parvizi, J., *et al.* (2000) Subcortical Brain Activity During the Feeling of Self-Generated Emotions, *Nature Neuroscience*, 3:10, pp. 1049–1054.

Damasio, A. R., Tranel, D. & Damasio, H. C. (1991) Somatic Markers and the Guidance of Behavior: Theory and preliminary testing, in: H. S. Levin, H. M. Eisenberg & A. L. Benton (eds), *Frontal Lobe and Dysfunction* (New York, Oxford University Press), pp. 217–229.

Davidson, R. J. (1999) Neuropsychological Perspectives on Affective Styles and their Cognitive Consequences, in: T. Dalgleish & M. Power (eds), *Handbook of Cognition* (Chichester, John Wiley & Sons), pp. 103–123.

Davidson, R., Jackson, D. & Kalin, N. (2000) Emotion, Plasticity, Context, and Regulation: Perspectives from affective neuroscience, *Psychological Bulletin*, 126, pp. 890–909.

Davis, M., Hitchcock, J. M. & Rosen, J. B. (1991) Neural Mechanisms of Fear Conditioning Measured with the Acoustic Startle Reflex, in: J. Madden IV (ed.), *Neurobiology, Learning, Emotion and Affect* (New York, Raven Press), pp. 67–95.

Dougherty, L. M., Abe, J. & Izard, C. E. (1996) Differential Emotions Theory and Emotional Development in Adulthood And Later Life, in: C. Magai & S. H. McFadden (eds), *Handbook of emotion, adult develoment, and aging* (San Diego, CA, Academic Press), pp. 27–41.

Ekman, P. (1993) Facial Expression and Emotion, *American Psychologist*, 48:4, pp. 394–392.

Ferrari, M. (2003) Baldwin's Two Developmental Resolutions of the Mind-Body Problem, *Developmental Review*, 23, pp. 79–108.

Forgas, J. P. (1999) Network Theories and Beyond, in: T. Dalgleish & M. Power (eds), *Handbook of Cognition and Emotion* (Chichester, John Wiley & Sons), pp. 591–611.

Geake, J. (2008) Neuromythologies in Education, *Educational Research*, 50:2, pp. 123–133.

Geary, D. (1996) Response: in: *Bridging the Gap between Neuroscience and Education*. Workshop Co-Sponsored by the Education Commission of the States and the Charles A. Dana Foundation, Denver, CO. 26–28 July.

Goleman, D. (1995) *Emotional intelligence* (New York, Bantam Books).

Goswami, U. (2004) Neuroscience and Education, *British Journal of Educational Psychology*, 74, pp. 1–14.

Gottman, J. M. & Katz, L. F. (2002) Children's Emotional Reactions to Stressful Parent-Child Interactions: The link between emotion regulation and vagal tone, *Marriage & Family Review*, 34:3, pp. 265–283.

Gray, J. A. (1991) Neural Systems, Emotion and Personality, in: J. Madden IV (ed.), *Neurobiology, Learning, Emotion and Affect* (New York, Raven Press), pp. 273–306.

Gray, J. A. (1999) Cognition, Emotion, Conscious Experience and the Brain, in: T. Dalgleish & M. Power (eds), *Handbook of Cognition and Emotion* (Chichester, John Wiley & Sons), pp. 83–102.

Gross, J. J. & John, O. P. (2003) Individual Differences in Two Emotion Regulation Processes: Implications for affect, relationships, and well-being, *Personality and Social Psychology*, 85:2, pp. 348–362.

Hall, J. (2005) *Neuroscience and Education: A review of the contribution of brain science to teaching and learning* No. 121 (Glasgow, University of Glasgow).

Hinson, J. M., Jameson, T. L. & Whitney, P. (2002) Somatic Markers, Working Memory, and Decision Making, *Cognitive, Affective, and Behavioral Science*, 2:4, pp. 341–353.

Izard, C. E. (1984) Emotion-cognition Relationships and Human Development, in: C. E. Izard, J. Kagan & R. B. Zajonc (eds), *Emotions, Cognition, and Behavior* (New York, Cambridge University Press), pp. 17–37.

Kagan, J. (1984) The Idea of Emotion in Human Development, in: C. E. Izard, J. Kagan & R. B. Zajonc (eds), *Emotions, Cognition, and Behavior* (New York, Cambridge University Press), pp. 38–72.

Lang, P. J., Davis, M. & Ohman, A. (2000) Fear and Anxiety: Animal models and human cognitive psychophysiology, *Journal of Affective Disorders*, 61:3, pp. 137.

Lazarus, R. (1994) Appraisal: The long and the short of it, in: P. Ekman & R. J. Davidson (eds), *The Nature of Emotion: Fundamental questions* (New York, Oxford University Press), pp. 208–215.

Lazarus, R. S. (1999) *Stress and Emotion: A new synthesis* (New York, Springer Publishing).

Ledoux, J. E. (1986) Sensory Systems and Emotion: A model of affective processing, *Integrative Psychiatry*, 4, pp. 237–248.

Ledoux, J. (1996) *The Emotional Brain: The mysterious underpinnings of emotional life* (New York, Simon & Schuster).

Ledoux, J. (2000) Cognitive-emotional Interactions: Listen to the brain, in: R. Lane & L. Nadel (eds), *The Cognitive Structure of Emotions* (New York, Oxford University Press), pp. 129–155.

Leventhal, H. (1982) The Integration of Emotion and Cognition: A view from the perceptual-motor theory of emotion, in: M. S. Clarke & S. T. Fiske (eds), *Affect and Cognition: The Seventeenth Annual Carnegie Symposium on Cognition* (Hillsdale, NJ, Lawrence Erlbaum Associates), pp. 121–156.

Lewis, M. D. & Steiben, J. (2004) Emotion Regulation in the Brain: Conceptual issues, *Child Development*, 75:2, pp. 371–376.

Maier, S. F. (1991) Stressor Controllability, Cognition and Fear, in: J. Madden IV (Ed.), *Neurobiology, Learning, Emotion and Affect* (New York, Raven Press), pp. 155–193.

Meinhardt, J. & Pekrun, R. (2003) Attentional Resource Allocation to Emotional Events: An ERP study, *Cognition and Emotion*, 17:3, pp. 477–500.

Moore, B., Underwood, B. & Rosenhan, D. L. (1984) Emotion, Self, and Others, in: C. E. Izard, J. Kagan & R. B. Zajonc (eds), *Emotions, Cognition, and Behavior* (New York, Cambridge University Press), pp. 464–483.

Ohman, A. (1999) Distinguishing Unconscious from Conscious Emotinal Processes: Methodological considerations and theoretical implications, in: T. Dalgleish & M. Power (eds), *Handbook of Cognition and Emotion* (Chichester, John Wiley & Sons), pp. 321–352.

Panksepp, J. (1990) Gray Zones at the Emotion/Cognition Interface: A commentary, *Cognition and Emotion*, 4:3, pp. 289–302.

Panksepp, J. (1994) Proper Distinction Between Affective and Cognitive Process is Essential for Neuroscientific Progress, in: P. Ekman & R. J. Davidson (eds), *The Nature of Emotion: Fundamental questions* (New York, Oxford University Press), pp. 208–215.

Parkinson, B., Totterdall, P., Briner, R. B. & Reynolds, S. (1996) *Changing Moods: The psychology of mood and mood regulation* (Harlow, Addison-Wesley Longman).

Parrott, W. G. & Spackman, M. P. (2000) Emotion and Memory, in: M. Lewis & J. Haviland-Jones (eds), *The Handbook of Emotions* (New York, Guilford Press), pp. 476–490.

Patten, K. E. (2004) Neuropedagogy: Imagining the learning brain as emotive mind. Paper presented at the Imaginative Education Group Conference, July, Vancouver, BC, Canada.

Patten, K. E. (2008) Toward a Neuropedagogy of Emotion. PhD thesis, Simon Fraser University.

Phelps, E. A. (2006) Emotion and Cognition: Insights from studies of the human amygdala, *Annual Review of Psychology*, 57, p. 27.

Porges, S. W. (2001) The Polyvagal Theory: Phylogenetic substrates of a social nervous system, *Psychophysiology*, 42, pp. 123–146.

Sabini, J. & M. Silver. (2005) Ekman's Basic Emotions: Why not love and jealousy? *Cognition and Emotion*, 19:5, pp. 693–712.

Scherer, K. R. (2000) Psychological Models of Emotion, in: J. C. Borod (ed.), *The Neuropsychology of Emotion* (New York, Oxford University Press), pp. 137–162.

Shipman, K. L., Zeman, J. L. & Stegall, S. (2001) Regulating Emotionally Expressive Behavior: Implications of goals and social partner from middle childhood to adolescence, *Child Study Journal*, 31:4, pp. 249–268.

Silk, J. S. (2002) Emotion Regulation in the Daily Lives of Adolescents: Links to adolescent development. PhD thesis, Temple University.

Silvia, P. J. (2002) Self-awareness and Emotional Intensity, *Cognition and Emotion*, 16:2, pp. 195–216.
Tugade, M. M. & Fredrickson, B. L. (2004) Resilient Individuals use Positive Emotions to Bounce Back from Negative Emotional Experiences, *Journal of Social Psychology*, 86:2, pp. 320–333.
Zajonc, R. B. (1980) Feeling and Thinking: Preferences need no inferences, *American Psychologist*, 35, pp. 151–175.
Zajonc, R. B., Pietromonaco, P. & Bargh, J. (1982) Independence and Interaction of Affect and Cognition, in: M. S. Clark & S. T. Fiske (eds), *Affect and Cognition: The Seventeenth Annual Carnegie Symposium on Cognition* (Hillsdale, NJ, Lawrence Erlbaum Associates), pp. 211–227.

14
Implications of Affective and Social Neuroscience for Educational Theory

MARY HELEN IMMORDINO-YANG

Advances in Social and Affective Neuroscience: Bringing Neuroscientific Evidence to Inform Educational Theory

Anyone involved in raising and educating children, from parents to teachers to coaches, mentors and beyond, knows that social learning is a major force in children's development. Typical children watch and engage with other people, imitate these other people's actions (including mental actions and beliefs), and look to trusted adults and peers for emotional and other feedback on their behavior. They imagine how other people feel and think, and those thoughts in turn influence how they feel and think.

Interestingly, evidence from social and affective neuroscience is shedding new light on the neural underpinnings of such social processing, affective responses and their relation to learning. These new discoveries link body and mind, self and other, in ways that only poets have described in the past (Casebeer & Churchland, 2003). They dissolve traditional boundaries between nature and nurture in development (Immordino-Yang & Fischer, 2010), and underscore the importance of emotion in 'rational' learning and decision-making (Damasio, 2005; Haidt, 2001; Immordino-Yang & Damasio, 2007). The challenge now for educators is to reconcile the new neuroscientific findings with established educational theories, to discover how this new information can be used to improve teaching and learning.

Our Bodies, Our Minds; Our Cultures, Our Selves

Traditional Western views of the mind and body, such as that of Descartes, divorced high-level, rational thought from what were thought of as the basal, emotional, instinctual processes of the body (Damasio, 2005 [1994]). By contrast, recent work in affective and social neuroscience has revealed a new view of the mind. Far from divorcing emotions from thinking, this research collectively suggests that emotions, such as anger, fear, happiness and sadness, are cognitive and physiological processes that involve both the body and mind (Damasio *et al.*, 2000). As such, they utilize brain systems for body regulation (e.g. for blood pressure, heart rate, respiration, digestion) and sensation (e.g. for physical pain or pleasure, for stomach ache). They also influence brain systems for cognition, changing thought in characteristic ways—from the desire to seek revenge in anger, to the search for escape in fear, to the receptive openness to others in happiness, to the ruminating on lost people or objects in

Educational Neuroscience, First Edition. Edited by Kathryn E. Patten and Stephen R. Campbell.
Chapters © 2011 The Authors. Book compilation © 2011 Educational Philosophy and Theory/Blackwell Publishing Ltd.
Published 2011 by Blackwell Publishing Ltd.

sadness. In each case, the emotion can be played out on the face and body, a process that can be felt via neural systems for sensing and regulating the body, or the emotion can sometimes involve simulations of the body that do not leave the brain. And in each case, these feelings interact with other thoughts to change the mind in characteristic ways, and to help people learn from their experiences. Put simply, what affective neuroscience is revealing is that the mind is influenced by an interdependency of the body and brain; both the body and brain are involved, therefore, in learning (Immordino-Yang & Damasio, 2007).

Further, educators have long known that thinking and learning, as simultaneously cognitive and emotional processes, are not carried out in a vacuum, but in social and cultural contexts (Fischer & Bidell, 2006). A major part of how people make decisions has to do with their past social experiences, reputation and cultural history. Now, social neuroscience is revealing some of the basic biological mechanisms by which social learning takes place (Frith & Frith, 2007; Mitchell, 2008). According to current evidence, social processing and learning generally involve internalizing one's own subjective interpretations of other people's feelings and actions (Uddin *et al.*, 2007). We perceive and understand other people's feelings and actions in relation to our own beliefs and goals, and vicariously experience these feelings and actions as if they were our own (Immordino-Yang, 2008). Just as affective neuroscientific evidence links our bodies and minds in processes of emotion, social neuroscientific evidence links our own selves to the understanding of other people.

For example, how do we know that the atrocities committed on 9/11/2001 are wrong? And why do most Americans have such a difficult time understanding how the terrorists were able to carry out these actions? We automatically, albeit many times nonconsciously, imagine how the passengers on those planes must have felt, empathically experiencing both what they were thinking about and their emotions around these thoughts. For many, just thinking of the images of planes hitting buildings induces a fearful mindset with all its physiological manifestations, like a racing heart and anxious thoughts. Similarly, we have difficulty empathizing with the terrorists who brought down the planes, because the values, morals and emotions that motivated these men are so different from our own.

Human Nature, Human Nurture

From the perspective of affective neuroscience, the social emotions that motivated the terrorists, as well as those we experience when empathizing with the passengers, represent a uniquely human achievement, and one that is relevant to education: the ability to feel emotions and engage in actions about the vicariously experienced beliefs of another person. Social emotions and their associated thoughts and actions are biologically built but culturally shaped; they reflect our neuropsychological propensity to internalize the actions of others, but are interpreted in light of our own social, emotional and cognitive experiences. Put another way, human nature is to nurture and be nurtured. We act on our own accord but interpret and understand our choices by comparing them against the norms of our culture, learned through social, emotional and cognitive experiences.

As is the case for basic emotions, the neural processes for experiencing and interpreting these various choices are not independent from our bodies. Instead social emotions, though arguably a pinnacle human achievement, remain biologically grounded in our most basic physiological life-regulatory processing. The feeling of these emotions appears to modulate the neural systems that sense stomach ache and regulate blood chemistry, for example. Especially intriguing, these emotions also involve systems associated with visceral self-awareness that are related to consciousness. Quite literally, it appears that the ability to treat others as we would be treated relies on feeling the empathic welling in our throat or 'punch' in our gut—feeling these on the substrate of our own psychological and bodily selves and interpreting them in light of personal experience and cultural knowledge, including that provided by education.

For example, let us take an educationally relevant problem—why does a student solve a physics problem? The reasons are fundamentally emotional, and range from pleasing his parents, to the intrinsic reward of finding the solution, to avoiding punishment or the teacher's disapproval, to the desire to attend a good college. Each of these reasons involve an implicit or explicit social or emotional value judgment, as the student imagines how others would react to his behavior, or how it would feel to solve the problem. And how does the student solve the problem? To apply problem-solving skills usefully in physics, the student must first motivate and engage himself sufficiently, must recognize the type of problem that is before him, and must call up information and strategies that will steer him toward a correct solution. Emotion plays a critical role in all of these stages of problem solving, helping the student to evaluate, either consciously or non-consciously, which knowledge and skills are likely relevant, and which will lead to a correct solution, based on his past learning. As he begins thinking through the solution, he is emotionally evaluating whether each cognitive step is likely to bring him closer to a useful solution, or whether it seems to be leading him astray. From a neuropsychological perspective, the brain systems for emotion form the 'rudder' that steers his thinking toward the development and recruitment of an effective skill (Immordino-Yang & Damasio, 2007), in this case for the solving of physics problems. Through regulating and inciting attention (Posner & Rothbart, 2005), motivation, and evaluation of possible social and cognitive outcomes, emotion serves to facilitate the student's recruitment of brain networks that support the skills he is developing. Here we use the example of solving a hypothetical physics problem, but the same mechanisms would be at play in the solving of other sorts of problems too, such as in deciding how to help one's friend or how to vote in a presidential election.

Emotion (Body and Mind) in Educational Context

Schools are social contexts. Each school is a community that functions inside a broader culture, and the social and emotional experiences that children have as members of a school's culture will shape their cognitive learning (Rueda, 2006). Children's bodies, brains and minds are meaningful partners in learning. Each child builds on his or her biological predispositions, his or her 'nature', grappling with his or her own biological and psychological 'self' as a platform on which to understand the thoughts and actions of other people, both peers and teachers.

When understood in this way, we can appreciate that even the driest, most logical academic learning cannot be processed in a purely rational way. Instead, the student's body, brain and mind come together to produce cognition and emotion, which are subjectively intertwined as the student constructs culturally relevant knowledge and makes decisions about how to act and think.

Taken together, the neuroscientific evidence linking emotion, social processing, and self, suggests a new approach to understanding how children engage in academic skills, like reading and math. While skills like reading and math certainly have cognitive aspects, the reason why we engage in them, the importance we assign to them, the anxiety we feel around them, and the learning that we do about them, are driven by the neurological systems for emotion, social processing and self. Neuroscientific evidence suggests that we can no longer justify learning theories that dissociate the mind from the body, the self from social context. To learn, students empathically recognize the teacher's actions, thoughts and goals, a process that reflects each student's own social and cognitive experiences and preferences. For example, to learn how to do a math problem, the students in the class must understand the goal of the exercise, and be able to relate that goal to the teachers' actions and thoughts, as well as to their own skills and memories. Using their own experience as a platform, the students struggle to discern and reconstruct the teacher's oftentimes invisible mental actions in their own mind. This process is subjective, emotional, and grounded in each student's predispositions and personal history.

Affective and Social Neuroscience and Educational Theory: A Plan for the Future

Despite their obvious relevance to educational environments, for the findings described above to have their full impact, educators and neuroscientists need to debate the general principles that the findings reveal, in order to derive testable hypotheses for education. In bringing neuroscience to inform education, this updating of educational theory is often neglected. Many times, educators and neuroscientists alike, caught up in their zeal for new and exciting information and seeing the desperate need to improve education, overlook the importance of theory building. Take, for example, the Mozart Effect—a sound scientific finding relating spatial ability to music listening that was vastly misrepresented and misapplied as a learning tool (see Rauscher, Shaw & Ky, 1993). Attempting to directly move from brain research to educational innovation without passing through a theory-building stage limits the generalizability of the new tool, and is sometimes even dangerous for children (Hirsh-Pasek & Bruer, 2007).

For education to truly benefit from these neuroscientific findings in a durable, deep way, for the full implications to become apparent, educators must examine closely the theory on which good practice is built, to reconcile the new and exciting evidence with established educational models and philosophies. For example, affective and social aspects of development are generally considered in examining curricula intended for young children. Affective and social neuroscience findings suggest, however, that emotion and cognition, body and mind, work together in students of all ages. Future research and theory in education should attempt to understand how best to characterize and capitalize on the emotional and social dimensions of learning in older students, including adults, keeping in mind what is known of the biological underpinnings of these processes.

In conclusion, there is a revolution imminent in education. The past decade has seen unprecedented advances in scientists' understanding of the brain and mind, and new information about the brain is expanding the influence of cognitive neuroscience into the classroom. The neuroscientific findings from affective and social neuroscience in particular could have profound implications for education, eventually leading to innovations in practice and policy. To discover these, we must lay the findings on the table for theoretical and philosophical debate. Irrespective of their scientific value, the individual brain findings are powerful for education only insofar as they suggest changes to our general knowledge of how learning and development happen. This is the next frontier for educational neuroscience. Neuroscientists and educators must work together to produce the Holy Grail: new ways of understanding development that have practical implications for the design of learning environments.

References

Casebeer, W. D. & Churchland, P. S. (2003) The Neural Mechanisms of Moral Cognition: A multiple-aspect approach to moral judgment and decision-making, *Biology and Philosophy*, 18:1, pp. 169–194.

Damasio, A. R. (2005) [1994] *Descartes' Error: Emotion, reason and the human brain* (Harmondsworth, Penguin Books).

Damasio, A. R. (2005) The Neurobiological Grounding of Human Values, in: J. P. Changeux, A. R. Damasio, W. Singer & Y. Christen (eds), *Neurobiology of Human Values* (London, Springer Verlag).

Damasio, A. R., Grabowski, T. J., Bechara, A., Damasio, H., Ponto, L. L. B., Parvizi, J., *et al.* (2000) Subcortical and Cortical Brain Activity During the Feeling of Self-Generated Emotions, *Nature Neuroscience*, 3:10, pp. 1049–1056.

Fischer, K. W. & Bidell, T. (2006) Dynamic Development of Action and Thought, in: W. Damon & R. Lerner (eds), *Handbook of Child Psychology, Vol. 1: Theoretical Models of Human Development*, 6th edn. (Hoboken, NJ, John Wiley & Sons), pp. 313–399.

Frith, C. D. & Frith, U. (2007) Social Cognition in Humans, *Current Biology*, 17:16, pp. R724–R732.

Haidt, J. (2001) The Emotional Dog and its Rational Tail: A social intuitionist approach to moral judgment, *Psychological Review*, 108:4, pp. 814–834.

Hirsh-Pasek, K. & Bruer, J (2007) The Brain/Education Barrier, *Science*, 317, p. 1293.

Immordino-Yang, M. H. (2008) The Smoke Around Mirror Neurons: Goals as sociocultural and emotional organizers of perception and action in learning, *Mind, Brain, and Education*, 2:2, pp. 67–73.

Immordino-Yang, M. H. & Damasio, A. R. (2007) We Feel, Therefore We Learn: The relevance of affective and social neuroscience to education, *Mind, Brain and Education*, 1:1, pp. 3–10.

Immordino-Yang, M. H. & Fischer, K. W. (2010) Neuroscience Bases of Learning, in: E. Baker, B. McGaw & P. Peterson (eds), *International Encyclopedia of Education*, 3rd Edition, *Section on Learning and Cognition* (Oxford, Elsevier).

Mitchell, J. P. (2008) Contributions of Functional Neuroimaging to the Study of Social Cognition, *Current Directions in Psychological Science*, 17:2, pp. 142–146.

Posner, M. I. & Rothbart, M. K. (2005) Influencing Brain Networks: Implications for education, *Trends in Cognitive Sciences*, 9:3, pp. 99–103.

Rauscher, F. H., Shaw, G. L. & Ky, K. N. (1993) Music and Spatial Task Performance, *Nature*, 365, p. 611.

Rueda, R. (2006) Motivational and Cognitive Aspects of Culturally Accommodated Instruction: The case of reading comprehension, in: D. M. McInerney, M. Dowson & S. V. Etten (eds), *Effective Schools: Vol. 6: Research on sociocultural influences on motivation and learning* (Greenwich, CT, Information Age Publishing), pp. 135–158.

Uddin, L. Q., Iacoboni, M., Lange, C. & Keenan, J. P. (2007) The Self and Social Cognition: The role of cortical midline structures and mirror neurons, *Trends in Cognitive Sciences*, 11:4, pp. 153–157.

Index

Educational Neuroscience, First Edition. Edited by Kathryn E. Patten and Stephen R. Campbell.
Chapters © 2011 The Authors. Book compilation © 2011 Educational Philosophy and Theory/Blackwell Publishing Ltd.
Published 2011 by Blackwell Publishing Ltd.